Optimisation of Renewable Energy Powered Desalination Systems

A sustainable techno-economic and environmental consideration

Online at: https://doi.org/10.1088/978-0-7503-6154-5

Optimisation of Renewable Energy Powered Desalination Systems

A sustainable techno-economic and environmental consideration

Ewaoche John Okampo
Department of Electrical and Electronic Engineering Science, University of Johannesburg, Johannesburg, South Africa

Nnamdi Nwulu
Department of Electrical and Electronic Engineering Science, University of Johannesburg, Johannesburg, South Africa

Pitshou Ntambu Bokoro
Department of Electrical and Electronic Engineering Technology, University of Johannesburg, Johannesburg, South Africa

IOP Publishing, Bristol, UK

ISBN 978-0-7503-6154-5 (ebook)
ISBN 978-0-7503-6152-1 (print)
ISBN 978-0-7503-6155-2 (myPrint)
ISBN 978-0-7503-6153-8 (mobi)

DOI 10.1088/978-0-7503-6154-5

Version: 20231201

IOP ebooks

British Library Cataloguing-in-Publication Data: A catalogue record for this book is available from the British Library.

Published by IOP Publishing, wholly owned by The Institute of Physics, London

IOP Publishing, No.2 The Distillery, Glassfields, Avon Street, Bristol, BS2 0GR, UK

US Office: IOP Publishing, Inc., 190 North Independence Mall West, Suite 601, Philadelphia, PA 19106, USA

This book is dedicated to all those who are facing water scarcity around the world and to all those who are making efforts to solve the problem of inadequate potable water for humanity.

Contents

Part III Optimisation of the desalination system with renewable energy sources

Part IV Techno-economic and environmental consideration

Preface

The world stands at a crossroads of energy and water challenges, where the imperatives of sustainable development, resource conservation, and climate change mitigation intersect. In this context, integrating renewable energy sources with desalination systems emerges as a beacon of hope—a solution that not only addresses the pressing issues of water scarcity but also aligns with the global transition toward cleaner, more resilient energy systems.

'Optimisation of renewable energy-powered desalination systems: a sustainable techno-economic and environmental consideration' delves into the heart of this critical juncture. It is a culmination of rigorous research, multidisciplinary exploration, and forward-thinking analysis that seeks to illuminate the pathway toward a more sustainable future. Within these pages, the intricate web of factors influencing the synergy between renewable energy and desalination is unravelled, offering a comprehensive understanding of the challenges and opportunities ahead.

As we embark on this journey, we explore the intricate dance between technology and economics, dissecting the intricate technical facets of renewable energy-powered desalination while juxtaposing them with economic considerations. The complex relationship between cost, efficiency, scalability, and environmental impact is dissected to reveal the true potential of these integrated systems. Through thorough evaluation and meticulous analysis, the book guides readers toward optimal configurations that balance the energy–water nexus.

However, no consideration of renewable energy and desalination is complete without addressing the vital environmental context. The book takes a holistic approach, examining various technologies' intricate environmental implications and highlighting their role in shaping sustainable development paradigms. It is a call to action, urging policymakers, researchers, and practitioners to recognise the inherent responsibility of safeguarding our planet's delicate balance.

This book is a culmination of the collaborative efforts of experts and enthusiasts who recognise the potential of renewable energy-powered desalination as a transformative force. It seeks to provide a comprehensive resource for academics, industry professionals, and decision-makers, offering insights that inform and inspire change.

As the world grapples with its pressing challenges, the significance of this intersection between renewable energy and desalination cannot be overstated. It is a testament to human innovation and the capacity to forge new paths, even in the face of daunting odds. The journey you are about to embark upon is one of exploration, learning, and, above all, a commitment to shaping a more sustainable and equitable world.

Let this book be your guide, companion, and source of inspiration as we navigate the complexities of optimising renewable energy-powered desalination systems, always with an eye on the horizon of a brighter future.

Acknowledgments

We would like to acknowledge the school of postgraduate studies and the faculty of Engineering, University of Johannesburg for the support and financial commitment to making the writing of this book easier.

Author biographies

Ewaoche John Okampo

Dr Ewaoche John Okampo, at the time of publication of this book, is a postdoctoral research fellow at the University of Johannesburg. He has obtained a PhD in electrical and electronic engineering from the University of Johannesburg, South Africa (2021). His PhD research focus was on energy–water nexus. He holds a Bachelor of Engineering and a Master of Engineering in electrical and electronic engineering from Caritas University, Enugu (2010) and University of Port Harcourt, Nigeria (2017), respectively. He also holds a postgraduate diploma in education from the National Open University of Nigeria (2016), and he is on course to complete a second PhD in operations management. His research interests include engineering education, energy–water nexus, renewable energy systems, systems modelling and simulation, and artificial/computational intelligence for power system optimisation and reliability. In his just-starting research journey, he has published 15 articles, of which 13 are indexed journals, and he has served as a reviewer for many high-rated journals.

Nnamdi Nwulu

Professor Nnamdi Nwulu is the Chair Holder of South Africa/ Switzerland Bilateral Research Chair in Blockchain Technology, Professor of Sustainable Cyber-Physical Engineering Systems and the Director of the Centre for Cyber-Physical Food, Energy and Water Systems (CCP-FEWS) at the University of Johannesburg. Professor Nwulu is also the acting head of the School of Electrical Engineering, University of Johannesburg.

Professor Nwulu's research focuses on FEWS and the inherent nexus between all three. Specifically, his research focuses on applying FEW-N thinking, digital technologies, mathematical optimisation techniques, and machine learning algorithms in food, energy, and water systems. This research focus has produced several books, inventions, and over 200 publications, winning him the 2021 National Research Foundation's Research Excellence Award for Early Career/Emerging Researchers in engineering/technology. He was also a 2020/2021 TW Kambule-NSTF-South32 Award finalist in the Emerging Researcher category and won the 2020 University of Johannesburg Vice-Chancellors' most promising young researcher award. He is a member of the prestigious South African Young Academy of Science.

He is a professional engineer registered with the Engineering Council of South Africa, a senior member of the Institute of Electrical and Electronics Engineers (SMIEEE), a senior member of the South African Institute of Electrical Engineers

(SMSAIEE) and a Y-rated researcher by the National Research Foundation of South Africa.

Professor Nwulu is the Editor-in-Chief of the *Journal of Digital Food Energy and Water Systems* and Associate Editor of the *IET Renewable Power Generation* and the *African Journal of Science, Technology, Innovation and Development.*

Pitshou Ntambu Bokoro

Pitshou Ntambu Bokoro, PhD, is Associate Professor in the Department of Electrical and Electronic Engineering Technology, University of Johannesburg, South Africa. He holds a senior membership with the SMSAIEE as well as with the SMIEEE. He obtained a Bachelor's degree in electrical engineering from Durban University of Technology, a Master's degree in electrical engineering from the University of Johannesburg, and a PhD from the University of The Witwatersrand, Johannesburg. He has published over 120 research papers (which include over 62 journal articles and 65 conference papers) in high-quality journals and peer-reviewed conference proceedings. He authored a couple of book chapters in reputed books published by IGI Global and the Institution of Engineering and Technology. He serves as a specialist editor in Energy and Power Systems for the *SAIEE Africa Research Journal.* He has also performed several reviews for various prestigious journals, including *IEEE Transactions on Power Delivery*, *IEEE Access*, *International Journal of Electrical Power and Energy Systems*, *Advances in Electrical and Electronic Engineering*, *International Transactions on Electrical Energy Systems*, *Materials Today: Proceedings* (Elsevier), *Journal of Engineering Design and Technology*, *Engineering Failure Analysis* (Elsevier), *Power Engineering Letters*, *Measurement* (Elsevier), *Journal of Electrical Engineering and Technology*, *Journal of Energy in Southern Africa*, *Electric Power Systems Research Journal*, *Electronics* (MDPI), *Energies* (MDPI), *Microelectronics Reliability*, and *Sustainability* (MDPI). He has supervised to completion over 18 postgraduate students (which include master's and doctoral students). His major research interests include renewable energy systems; power systems; power system reliability; distributed generation; surge arresters; and insulation and dielectrics, power quality, and condition.

Part I

Background information

Chapter 1

General introduction

Modern societies rely on both energy and water as essential lifelines, driving economic growth, societal well-being, and environmental balance. However, the dual challenges of energy scarcity and water stress have cast a shadow over global sustainability efforts. Energy production, particularly in its traditional forms, demands substantial water inputs, while water supply and treatment rely heavily on energy-intensive processes. Climate change further exacerbates these challenges, amplifying the vulnerability of energy and water systems. The inseparable interdependence of energy and water resources has emerged as a critical nexus with far-reaching implications for sustainable development. This chapter introduces the challenges posed by the availability of energy and water resources, highlighting the significance of their intricate nexus. The importance of the energy–water nexus lies not only in acknowledging the challenges but also in recognising the opportunities it presents. Addressing these intertwined issues requires a holistic approach that considers both resources' synergies and trade-offs. Technological innovations, such as renewable energy integration, smart grid systems, and water-efficient technologies, offer pathways to mitigate the impact on resource availability and enhance overall system resilience.

1.1 Introduction

It is a prominent and well-known fact that energy and water are critical components of a healthy economy. We all need these resources for daily living. Hence, the importance of these two resources cannot be overemphasised. Despite the global abundance of energy and water in their natural forms, there is still a high level of scarcity of processed or usable forms of these resources in many parts of the world. For this reason, international organisations like the United Nations propose and pursue sustainable development goals (SDGs) 6 and 7, which recommends 'clean

doi:10.1088/978-0-7503-6154-5ch1

water and sanitation' and 'affordable and clean energy for all,' respectively. As lofty as the ideas of the SDGs are, the failure in many governments of different nations to provide their citizens with affordable, reliable, sustainable clean energy and water has been well documented. This failure has, over time, been connected to several issues, such as climate change, population growth, stress on resources, and waste caused by human activities. To tackle these challenges, governments, researchers, and industries have, over the years, continued to develop and embrace technologies that focus on the production and efficient utilisation of energy and water. In doing so, they face further challenges, including high cost of production and environmental hazards. Desalination, for example, is at the forefront as one of the most effective means of freshwater production but is considered energy-intensive and expensive with the additional challenge of brine management. In the same vein, renewable energy sources are regarded as alternative solutions to the problem of carbon emissions from conventional energy sources. Still, the irradiation and inconsistency of some renewable energy sources remain as big limitations.

There is a current surge in research and ongoing conversation towards finding effective and efficient utilisation of a possible nexus of critical resources like energy, water, food, and environment to explore relative advantage. In this regard, several opinions and research outputs have already been documented. Still, most can only capture and tackle one or a few of the complex challenges that must be overcome to provide adequate, affordable, reliable, and sustainable clean energy and water. Holistic techniques are needed to achieve a more comprehensive solution to these multifaceted, complex problems [1]. Therefore, this book explores classical optimisation concepts to design techno-economic models for energy-efficient desalination systems with minimal or zero brine disposal to the environment. The book conceptualises desalination techniques and focuses mainly on the most common method, the reverse osmosis desalination technique. Explore the merit of this technique and its hybridisation with other approaches to achieve optimal freshwater production at minimal cost and environmental impact. This book also presents different mixed scenarios for energy-efficient management needed for water treatment. This led to the design of optimal models with renewable energy sources in the energy mix to reduce environmental pollution of carbon emission of convectional energy sources while complementing the intermittent limitation of some of the renewable energy sources. Compared to other similar books, the uniqueness of this book is its holistic approach in addressing most of the significant challenges hampering the delivery of adequate, affordable, reliable, sustainable clean energy and water.

This book is essential for government water board agencies to guide policy formulation to provide potable water at affordable prices with minimal or zero environmental impact. It could also serve as a techno-economic model for desalination and other water treatment industries for efficient energy and brine management during freshwater production. Furthermore, this book provides fundamental literature for researchers and academics seeking to expand the frontiers of knowledge in energy optimisation and freshwater production, as it contains different optimisation models that could serve as a foundation for future research. It could be used by undergraduate and postgraduate electrical and electronic engineering students as a handbook for energy-efficient management, application of optimisation techniques, and environmental sustainability models.

1.1.1 Background

The abundance of energy and water is yet to transcend into adequate use, mainly because of infrastructural deficit, and inefficient management and poor utilisation of both resources and existing infrastructures. These resources are essential for human existence, especially in their usable form.

The concept of the energy–water nexus defines the interrelationship between the two resources to maximise synergies and address trade-offs [2]. Due to the apparent significance of the two commodities (energy and water) for the human population, this idea has gained significant prominence in recent years. This mutual relationship between energy and water could be used in two ways: water utilisation for energy generation or energy expended for water purification. It could also be the case of both energy and water interdependence for cogeneration. In any case, the energy–water nexus is beneficial for domestic utilisation, industrial use, and agricultural efficiency through irrigation for food production. The importance of energy and water as independent commodities and other beneficial resources such as food has made understanding the nexus more important. This is because the nexus ensures the sustainability of both resources to the extent of their dependent advantage resources. Therefore, the energy–water nexus improves the production capacity and enhances the availability of these important resources needed for continuous human survival.

With constant global population growth, the demand for energy and irrigation increases exponentially, putting stress on the current infrastructure designed to ensure their continuous supply. In recent years, the energy–water nexus has received a lot of attention and research. This has resulted in commendable solutions to various problems associated with the nexus; however, the trade-offs between integration leave room for further study.

According to [3], 'there are five different drivers on which the trend in the nexus studies is based. These factors include (i) an increase in demand for both resources because of population growth; (ii) the economic increase in per capita demand for both resources; (iii) the degradation of available freshwater due to climate change and concern for negative environmental factors, and (iv) electric power and water-intensive electricity as a result of ageing infrastructure. These factors exert strain on the energy–water nexus.'

1.1.2 Freshwater accessibility

It is estimated that two-thirds of the earth's surface is covered by water, but only 1% of this water is suitable for domestic and industrial use [4]. In the past decade, one-fifth of the world's population has experienced severe water shortage [5]. One-fourth of the global population has access to water sources, but inadequate purification facilities remain a major problem [5]. According to the United Nations World Assessment Programme [6], by 2030 up to 40% of the global population will be affected by water scarcity. Currently, 1.3 billion people—including 450 million children—live in regions with extremely high water vulnerability [7]. The main causes of insufficient freshwater resources are climate change and pollution [8]. According to the World Health Organisation report titled 'State of the world's drinking water: an urgent call to action to accelerate progress on ensuring safe drinking water for all' [9], other factors that threaten water sustainability include flooding, which disrupts supplies, and pollution, which threatens both human health and ecosystems as a whole. Furthermore, urbanisation and population growth impede water to millions of people living in informal communities and settlements. In contrast, contaminated water sources are an everyday occurrence in rural areas. With 97% of all available water resources consisting of salt water, desalination and water reuse technologies are the obvious means for obtaining sustainable freshwater [8], [10]. However, the greatest obstacle to the implementation of these technologies is their high energy demand [11].

1.1.3 Energy accessibility

Given the current state of the world's energy, it is estimated that 18% of the world's population lacks access to electricity [12–14]. According to the International Energy Agency, global total electricity consumption in 2016 amounted to 20 863 TWh, with an annual rise of at least 3.2% projected based on the increase in the previous year. Coal supplied 38.3% of this total, followed by natural gas at 23.1%, hydroelectric power at 16.6%, and nuclear power at 10.4%, while solar/wind/geothermal/tide supplied 5.6%, oil at 3.7%, and biofuel and waste at 2.3% [15]. Only 34.9% of the total energy supply mix was renewable, with hydroelectric power accounting for the highest percentage. Both coal and natural gas are substantial contributors to climate change due to the quantum of carbon dioxide emitted per unit of energy produced [15]. However, coal is responsible for the highest percentage of the global energy supply mix, as it releases approximately twice as much CO_2 as natural gas. According to the report on the progress of sustainable development energy [16], achieving universal access to affordable, reliable, sustainable, and modern energy by 2030 is still an ambitious goal. This can only be achieved if there is a significant increase in the share of renewable energy in total final energy consumption, which stood at 17.7% in 2019.

In addition, the figures on the usage of generated energy show that around 40% is used for residential purposes, 47% is used for industrial purposes, and 13% is lost during the transmission [15]. As a result of the harmful effects that conventional energy sources derived from fossil fuels have on the environment—including greenhouse gas emissions and pollution—renewable energy sources (RES) are the only realistic choice that should be taken into consideration [17, 18]. In their own right, RES have several significant drawbacks, such as variable sun irradiation and variable wind speed; hybridisation is typically used to circumvent these problems [19].

1.1.4 Merit of the nexus

There is a direct relationship between energy and water, so that water is used for the generation of energy in the form of salinity gradient and hydro-energy generation, while energy is needed for water purification in the process of desalination or water reuse. Therefore, a water crisis can lead to an energy crisis and vice versa. These resources will dominate human life in the future, as they are critical to civilisation and industrialisation [20]. There are concerns about their cost of production and impact on the environment. The nexus of these two essential resources seeks to maximise synergy and address the trade-offs between the two. Therefore, some of the merits of this nexus include:

(a) harmonisation and reduction of production cost for both resources;
(b) minimisation or total elimination of carbon footprint from fossil fuel energy sources;
(c) minimisation of the negative environmental impacts of brine disposal, as the brine can be reused for renewable energy generation;
(d) improvement of energy and water-efficient utilisation and management;
(e) increasing the productivity and availability of both resources;
(f) increasing potential for food production; therefore, we reduce hunger and poverty;
(g) in general, the energy–water nexus has the potential to become a critical foundation for urbanisation, industrialisation, and environmental sustainability.

1.2 Summary

Water scarcity is a severe issue in many parts of the world. Although it is an energy- and cost-intensive process, desalination has been identified as a viable option for freshwater production by capitalising on the seawater abundance. There are environmental costs associated with this method as a result of the disposal of brine. Some of the most popular renewable energy sources are limited by their susceptibility to interruption or irradiation, and the environmental impacts of carbon emissions from conventional energy sources are discouraging. Therefore, this book presents several mathematical models of optimal energy sources, incorporating a demand response programme for an efficient energy schedule to power a reverse osmosis desalination system at a minimal levelized cost of energy. This is an effort to address most of the significant challenges hindering the delivery of adequate, affordable, reliable, sustainable clean energy and water. Additionally, a mathematical optimisation model that considers carbon emission and the cost of maximising freshwater production in a reverse osmosis desalination system is presented. Furthermore, a techno-economic optimisation model of a combined desalination and brine treatment unit has been developed for freshwater generation and brine management.

References

[1] Okampo E J and Nwulu N 2021 Optimisation of renewable energy powered reverse osmosis desalination systems: a state-of-the-art review *Renew. Sustain. Energy Rev.* **140** 110712

[2] Laspidou C *et al* 2018 Energy modelling and the Nexus concept *Energy Strateg. Rev.* **19** 1–6

[3] Lubega W N and Farid A M 2016 A reference system architecture for the energy–water nexus *IEEE Syst. J.* **10** 106–16

[4] Abdelkareem M A, El Haj Assad M, Sayed E T and Soudan B 2018 Recent progress in the use of renewable energy sources to power water desalination plants *Desalination* **435** 97–113

[5] Jian-yun Z, Man-ting S, Rui-min H, Shahid S, Xing-hui X and Xiao-jun W 2014 Catastrophe theory to assess water security and adaptation strategy in the context of environmental change *Mitig. Adapt. Strateg. Glob. Chang.* **19** 463–77

[6] UNESCO 2015 Water for a sustainable world *The United Nations World Water Development Report 2015*

[7] UNICEF 2021 Water security for all *Reimagining WASH*

[8] Prisciandaro M, Capocelli M, Piemonte V and Barba D 2016 Process analysis applied to water reuse for a 'closed water cycle' approach *Chem. Eng. J.* **304** 602–8

[9] WHOUNICEFWorld Bank Group 2022 *State of the World's Drinking Water: An Urgent Call to Action to Accelerate Progress in Ensuring Safe Drinking Water For All*

[10] Ziolkowska J R and Reyes R 2016 Impact of socio-economic growth on desalination in the United States *J. Environ. Manage.* **167** 15–22

[11] Okampo E J and Nwulu N 2022 Economic and reliability assessment of hybrid PRO-RO desalination systems using brine for salinity gradient energy production *Sustainability* **14** 3328

[12] Maleki A, Khajeh M G and Rosen M A 2016 Weather forecasting for optimisation of a hybrid solar-wind–powered reverse osmosis water desalination system using a novel optimiser approach *Energy* **114** 1120–34

[13] Maleki A, Pourfayaz F and Ahmadi M H 2016 Design of a cost-effective wind/photovoltaic/hydrogen energy system for supplying a desalination unit by a heuristic approach *Sol. Energy* **139** 666–75

[14] Maleki A 2018 Design and optimization of autonomous solar-wind-reverse osmosis desalination systems coupling battery and hydrogen energy storage by an improved bee algorithm *Desalination* **435** 221–34

[15] Marletta L, Evola G, De Simone M, Bruno R, Oliveti G and Arcuri N 2014 Solar energy *Building Refurbishment for Energy Performance* ed A Magrini (Switzerland: Springer Cham) pp 159–214

[16] IEAIRENAUNSDWorld BankWHO 2022 *Tracking SDG 7: The Energy Progress Report*

[17] Okampo E J and Nwulu N 2021 Techno-economic evaluation of reverse osmosis desalination system considering emission cost and demand response *Sustain. Energy Technol. Assess.* **46** 101252

[18] Okampo E J and Nwulu N I 2020 Optimal energy mix for a reverse osmosis desalination unit considering demand response osmosis *J. Eng. Des. Technol.* **18** 1287–303

[19] Okampo E J and Nwulu N 2020 Optimal design and techno-economic evaluation of renewable energy powered combined reverse osmosis desalination and brine treatment unit *Desalin. Water Treat.* **202** 27–37

[20] Razeghi M, Hajinezhad A, Naseri A, Noorollahi Y and Moosavian S F 2023 Multi-criteria decision for selecting a wind farm location to supply energy to reverse osmosis devices and produce freshwater using GIS in Iran *Energy Strateg. Rev.* **45** 101018

Part II

Understanding water treatment techniques

Chapter 2

Water treatment techniques

Water treatment techniques play a pivotal role in alleviating its scarcity by ensuring the provision of clean and safe drinking water. This chapter presents an overview of various advanced water treatment methods, their mechanisms, and their applications with emphasis on desalination techniques. Desalination has been identified as a leading water treatment technology capable of addressing the global challenge of water scarcity by providing access to freshwater resources from saline water bodies. The study offers an in-depth exploration of desalination technologies, encompassing both thermal and membrane-based processes. It navigates the complexities of distillation techniques, including multi-effect distillation and multi-stage flash distillation, elucidating their principles of vaporisation and condensation. Additionally, reverse osmosis, electrodialysis, and forward osmosis as membrane-based methods were discussed, highlighting the role of semi-permeable membranes in separating salt ions from water molecules. Reverse osmosis is identified as the commonly used desalination method because of its lower energy requirement and simple but effective configuration. It is recommended that this method be combine with other methods such as forward osmosis and nanofiltration methods for better performance. The hybrid methods enhance the quality of the desalinated water and have a brighter chance at longevity and sustainability.

2.1 Wastewater treatment and water reuse technology

The use of wastewater and water reuse techniques is an approach to the problem of freshwater sustainability. Wastewater refers to any contaminated water used for domestic, commercial, industrial, or agricultural purposes. Wastewater can contain various contaminants, such as organic matter, nutrients, heavy metals, and pathogens, making it potentially hazardous to human health and the environment. However, by treating and reusing wastewater, we can effectively increase the

doi:10.1088/978-0-7503-6154-5ch2

quantity of available water while decreasing contamination and improving public health [1]. Implementing wastewater and water reuse techniques to combat water scarcity has additional benefits. These include a reduction in water consumption, an improvement in water quality, and an increase in available water for agricultural and industrial purposes. Additionally, water reuse can reduce energy consumption and greenhouse gas emissions associated with freshwater extraction and treatment. When implementing these techniques, one must also consider the cost of infrastructure and treatment systems, potential public health concerns, and regulatory and policy barriers. Reuse of water must be carefully planned and managed to be safe, effective, and sustainable over the long term. Before water can be reliably reused, physical, chemical, and biological treatments are frequently used to remove contaminants and pathogens.

Physical treatment, also known as primary treatment methods, is the first stage of wastewater treatment and consists of the removal of suspended solids and other residues [2]. This is achieved by utilising a series of screens, grit chambers, and sedimentation containers. The screens remove large objects such as twigs, rags, and plastics. The grit chamber removes sand, gravel, and other heavy substances that could damage compressors and machinery. In sedimentation tanks, wastewater is held for a period to allow heavier solids to settle to the bottom and lighter substances to float to the surface, where they are skimmed off.

Physical/primary treatment methods have the following advantages [3]:

- Simple and inexpensive process
- Less energy consumption than other treatment methods
- Easy to operate and maintain
- Effectively removes larger particles and detritus

Physical/primary treatment methods have the following disadvantages:

- Does not remove all pollutants, especially dissolved contaminants
- May generate odours and sludge that require additional treatment and disposal

Biological and chemical methods comprise the secondary remediation process. Utilising microorganisms, biological remediation methods decompose organic matter in wastewater [4]. Typically, this method uses aeration containers and biological filters. This method uses microorganisms to decompose organic matter and eliminate pollutants. The most prevalent secondary treatment technique is activated sludge treatment [3]. This procedure combines wastewater and bacterial culture in an aeration tank. In the aeration tanks, oxygen is supplied to encourage the development of bacteria that digest organic matter. The water is then disinfected with chlorine or ultraviolet radiation before being released into the environment. The bacteria decompose organic matter and generate sludge, accumulating in a sedimentation tank. Subsequently, the purified water is disinfected before discharge. Using chemicals, chemical treatment methods remove pollutants from the effluent. This procedure typically employs coagulants and disinfectants. Coagulants are chemicals that make suspended solids and other contaminants easier to remove by causing them to clump together. Common coagulants include aluminium sulphate and ferric chloride [5]. Disinfectants are chemicals that destroy microorganisms and bacteria in wastewater. The most common disinfectants are chlorine and ozone.

Secondary treatment methods have the following advantages:

- Secondary treatment is effective in removing smaller contaminants and particles.
- It is capable of disinfecting water.
- It eliminates a significant amount of organic matter and pollutants.
- It can be combined with primary treatment to create a more comprehensive strategy.
- Less sediment is produced than with primary treatment.
- It is capable of removing both dissolved and suspended organic matter.
- It generates less sludge than tertiary treatment.
- It enhances the quality of the water that is discharged into the environment.

Disadvantages of secondary treatment include the following:

- Secondary treatment requires a more significant energy input than primary treatment.
- It may require additional nitrogen and phosphorus removal treatment.
- It can be sensitive to temperature and pH fluctuations.

The final stage of wastewater treatment (tertiary treatment) removes any remaining contaminants that were not removed in the primary or secondary phases. This procedure involves filtration, disinfection, and the elimination of nutrients. The wastewater is passed through a series of filters, such as sand or activated carbon filters, to remove any remaining impurities. The remaining microorganisms and viruses are then eliminated using chlorine or ultraviolet light. To prevent eutrophication in receiving waters, nutrient removal involves removing excess nitrogen and phosphorous from wastewater.

Tertiary treatment has the following benefits:

- Produces high-quality water for reuse.
- Prevents environmental degradation.
- Extremely efficient in eliminating contaminants from wastewater.
- Enhances the quality of water to meet strict discharge standards.
- Appropriate for nonpotable water recycling.

After the primary and secondary treatment processes have been completed, advanced wastewater treatment methods are utilised to purify wastewater further. These techniques are intended to remove any residual contaminants from wastewater and produce a high-quality effluent that can be released into the environment or reused for nonpotable purposes [6]. The use of additional processes to remove specific contaminants from wastewater constitutes advanced treatment. These processes include membrane filtration, reverse osmosis, and ion exchange [7]. Membrane filtration removes impurities at the molecular level by passing effluent through a series of membranes [8]. Reverse osmosis removes dissolved solids and salts from effluent by forcing it through a semi-permeable membrane utilising high pressure. Ion exchange removes specific contaminants by exchanging ions in effluent with ions in a resin. Several advanced wastewater treatment methods can be implemented depending on the specific demands of the wastewater treatment plant and the regulatory requirements. Some of the most common advanced treatment methods include the following:

Chemical precipitation: Chemical precipitation is the addition of chemicals to effluents, such as alum or ferric chloride, to precipitate out dissolved solids and suspended particles. The precipitates are separated from the water using settling containers or flotation devices. Chemical precipitation effectively removes nutrients such as phosphorus and nitrogen that primary or secondary treatment procedures cannot remove.

Biological nutrient removal: Biological nutrient removal is a procedure that uses microorganisms to remove nutrients from wastewater, such as nitrogen and phosphorus [9]. The microorganisms are added to the wastewater and allowed to grow under controlled conditions in specialised tanks. Microorganisms eliminate nutrients by converting them into harmless gases like nitrogen and oxygen. Biological nutrient removal is a highly efficient method of removing nutrients from effluent and is often combined with secondary treatment.

Ozonation: Ozonation is a procedure in which ozone gas is used to disinfect and oxidise wastewater [3]. Ozone gas is produced by passing air through a high-voltage electrical discharge. The ozone gas is then combined with the contaminated water and allowed to react with the contaminants [10]. The removal of bacteria, viruses, and other pathogens from effluent by ozonation is extremely efficient [11]. In addition, it is effective at removing organic and odour-causing compounds from water.

Membrane filtration: Membrane filtration is the separation of suspended solids and dissolved particles from effluent using a semi-permeable membrane. Microfiltration, ultrafiltration, nanofiltration, and reverse osmosis are frequently used membrane filtration techniques [8]. These techniques eliminate contaminants from wastewater, such as bacteria, viruses, suspended solids, and dissolved organic matter [8]. Membrane filtration is a highly effective method of purifying wastewater that can generate the highest quality effluent. However, permeability is obstructed by membrane fouling, resulting in additional operation costs due to energy consumption and regular chemical cleaning [12]. This significant drawback is the main disadvantage of this technique. Consequently, integrated membrane processes can provide the optimal solution with respect to separation efficiency, performance, fouling control, and cost [13]. Multiple-step or integrated processes that combine biological, electrochemical, and membrane-based technologies are required to enhance the performance and purity of water products [14].

Advantages of advanced wastewater treatment:

- It produces high-quality effluent that is suitable for reuse or release into the environment.
- It removes a broad spectrum of contaminants, including dissolved solids, dispersed solids, nutrients, and pathogens.
- It reduces the environmental impact of wastewater discharge.

- It enhances the overall efficacy of the wastewater treatment process.
- It is capable of removing specific contaminants that cannot be eliminated in other treatment stages.
- It can produce high-quality water for reuse in specific industries, such as semiconductor manufacturing and pharmaceuticals.
- It can be combined with other treatment methods for improved efficacy.

The comparative benefits of wastewater treatment techniques are correlative. Several variables, including the nature of the wastewater, the discharge standards, and the local regulations, influence the selection of a wastewater treatment method. Although primary treatment is easy and inexpensive, it is ineffective in removing dissolved solids and nutrients. It efficiently removes organic matter and nutrients but consumes more energy and generates more sediment. Tertiary treatment is highly effective in removing contaminants, but its implementation is more costly. Treating wastewater is crucial for the protection of the environment and public health. Several factors influence the choice of treatment technique, and a combination of methods can be used to achieve the desired water quality.

2.2 Desalination: overview, methods, and configurations

Desalination is the purification of saline into potable water for domestic and industrial use. The excessive salt content of seawater and brackish water makes them unsuitable for drinking water or for industrial purposes. These varieties of water are referred to as saline water, which accounts for the greatest quantity of available water. Only 1% of the available water is fit for human consumption, for either domestic or industrial use [14]. Desalination is therefore regarded as a potent method of obtaining freshwater to address water scarcity and rising demand [15]. Recent interest in desalination as a viable option for freshwater production demonstrates its importance as a viable option. This has led to the development of various processes classified as thermal or membrane processes [14].

Some common examples of thermal desalination methods are as follows:

(a) Humidification-dehumidification (HDH)
(b) Multi-stage flash (MSF) distillation
(c) Vapour compressor distillation (VCD)
(d) Multi-effect distillation (MED)

These methods are based on the change of state, while membrane techniques use membranes with unique chemical and physical properties, such as ion adsorption and desorption, ion selectivity, and semi-permeability for separation of salt from water. Some of the commonly used membrane methods include:

(a) Capacitive deionisation (CDI)
(b) Electrodialysis (ED)
(c) Reverse osmosis (RO)
(d) Electrodialysis reversal (EDR)
(e) Membrane distillation (MD)

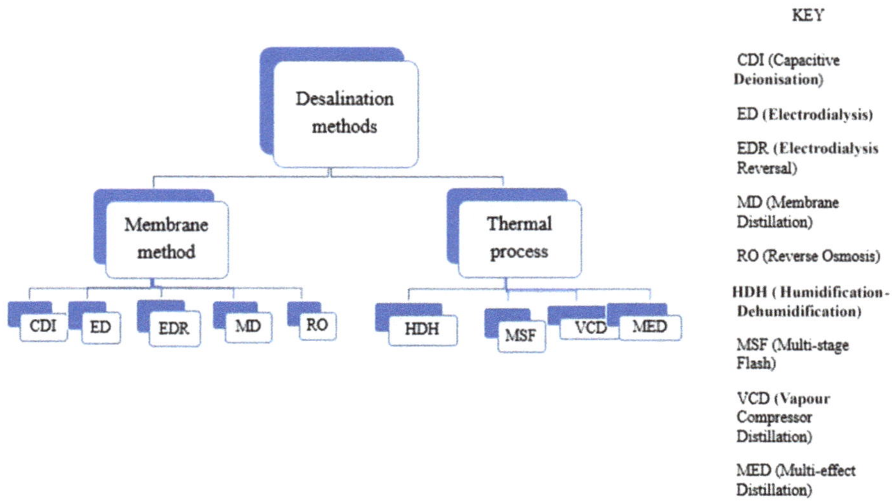

Figure 2.1. Classification of the main desalination methods. Reproduced from [19], copyright (2021) with permission from Elsevier.

Figure 2.1 explicitly depicts the classification of the desalination process and various techniques.

Other desalination methods include forward osmosis (FO), nanofiltration (NF), freezing or crystallisation technique, and hybrid methods. Over the years, RO, MSF, and MED have been the most frequently used methods among the various desalination methods. Figure 2.2 shows the global installation capacities of the desalination methods over different periods.

The operational principles of different desalination methods are presented in the following subsections.

2.2.1 Humidification-dehumidification technique

The HDH desalination method is based on the natural water cycle principle of evaporation and condensation [18]. In the humidification tower or humidifier, saline water is heated and the relative humidity of the ambient air increases to saturation based on the design temperature specification, resulting in water evaporation [19]. By increasing its relative humidity, air transports water from a humidifier to a dehumidifier, where the condenser is used to lower the air temperature and condense freshwater [20]. Based on the cycle configuration, HDH is classified as closed-air open-water, closed-air open-air, and open-air open-water [21]. It could also be heated by air or water [19, 22]. This method of desalination requires inexpensive materials for its simple configuration and can utilise low-grade thermal energy to reduce the carbon footprint [19]. Compared to other desalination techniques, such as MED and MSF, it requires a greater amount of thermal energy per unit of freshwater produced. Another limitation is the difficulty scaling up the energy

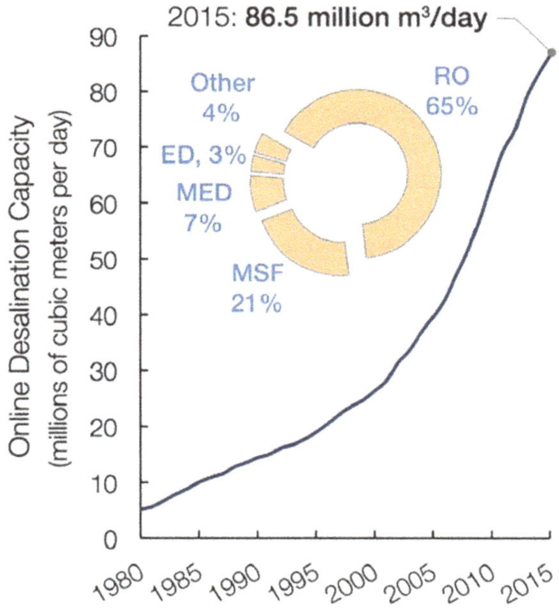

Figure 2.2(a). Total desalination unit capacity (worldwide) 1980–2015 [14, 16]. Reprinted from [16], copyright (2021) with permission from Elsevier.

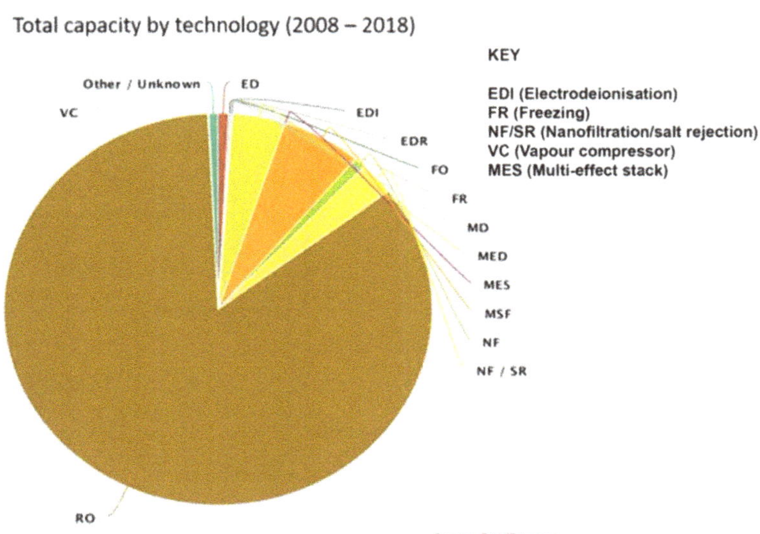

Figure 2.2(b). Global installation capacity of desalination methods 2010–2018 [16, 17]. Reprinted from [16], copyright (2021) with permission from Elsevier.

requirement [19, 22]. As a result of air circulation transporting evaporated water or vapour from the humidifier to the dehumidifier, some improved HDH configurations seek to reduce specific energy consumption (SEC). Müller-Holst [23, 24] proposed the method of multiple-effect humidification utilising innovative geometry. This configuration reduces and maintains a deficient energy requirement by inserting parallel plates as high-performance film condensers, which improves the continuous temperature distribution and natural air circulation. The authors present the incorporation of absorption beds to improve the close-water open-air cycle configuration of HDH to HDH adsorption (HDHA) [18]. This configuration allows saturated air at the lowest temperature to be sent to the absorption beds for dehydration, where the humidity is reduced through absorption, and dried air to be sent to the humidifier. The evaporation of the saturated air at the bottom allows for greater distillate production per cycle. The authors [25] develop a model of the configuration of the HDH cycle based on the first principle of thermodynamics, demonstrating heat and mass transfer in the HDH heat pump cycle. The HDH system uses a heat pump as a source of heating and cooling to increase the output ratio and energy recovery of the system. Figure 2.3 presents a schematic diagram of a combined water-heated closed-air open-water HDH cycle and heat pump, where a, w, and r represent air, water, and refrigerant, respectively, while 1, 2, 3… are state points.

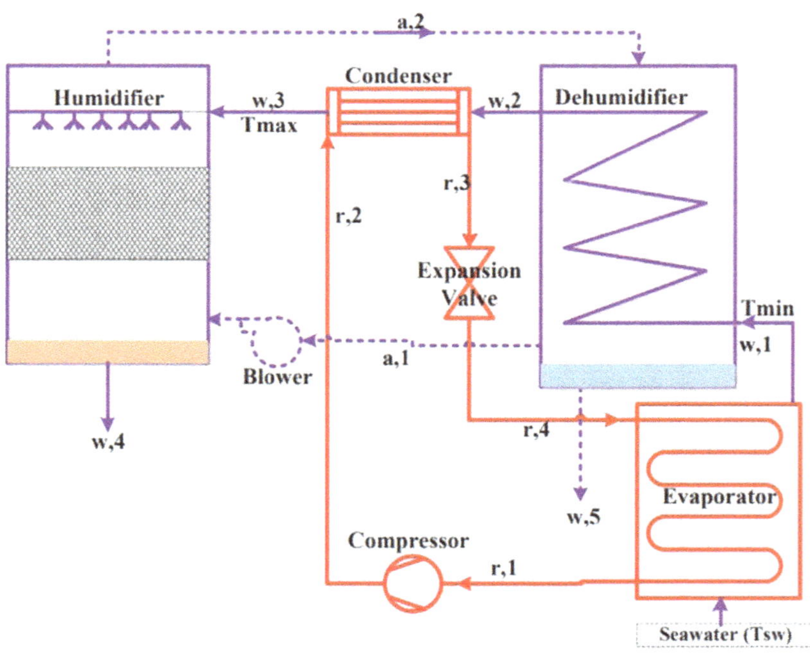

Figure 2.3. Schematic diagram of a combined water-heated, close-air open-water HDH cycle and heat pump [16, 25]. Reprinted from [16], copyright (2021) with permission from Elsevier.

2.2.2 Multi-stage flash technique

The brine heater, the heat rejection section, and the heat recovery section are the three sections that are typically included in an MSF plant layout. The MSF desalination method requires heating the brine, performing flash distillation in many phases, and eventually recovering the heat [26]. According to the authors [27], the hot brine is pumped into a pool of liquid, which is then used for evaporation. Because the temperature of the tube's surface is lower than that of the vapour, it causes the vapour to flash from the pool towards the tube bundles and then condense on the surface of the tube. Freshwater resulting from the vapour condensation is collected while heated brine continues to the next stage. Condenser tubes circulate low-temperature saltwater (brine) with the purpose of absorbing latent heat from the vapour formed by the high-temperature brine pool during this operation. The vapour is produced by the high-temperature brine pool. The process of heat recovery begins when the latent heat that has been absorbed begins to partially heat the brine that is circulating at a low temperature. This helps to save energy while also improving the efficiency of the system. MSF can be designed in several distinct configurations, the most common of which are brine recirculation (MSF-BR), the once-through (MSF-OT), and simple mixer (MSF-M) designs. MSF-BR is the most effective of these configurations, while the MSF-OT has the simplest design [24]. The optimisation of various MSF designs has been the subject of investigation in a lot of research. The authors in [28–30] detail the application of MSF-BR, while the authors in [31, 32] present MSF-M and MSF-OT, respectively. In reference [33], the authors offer a method to improve the resistance to corrosion in MSF desalination plants during an acid cleaning operation by employing a cationic surfactant. On the other hand, reference [34] examines the impact of various MSF facilities on the environment and human health using a life cycle evaluation. Figure 2.4 shows a general MSF process.

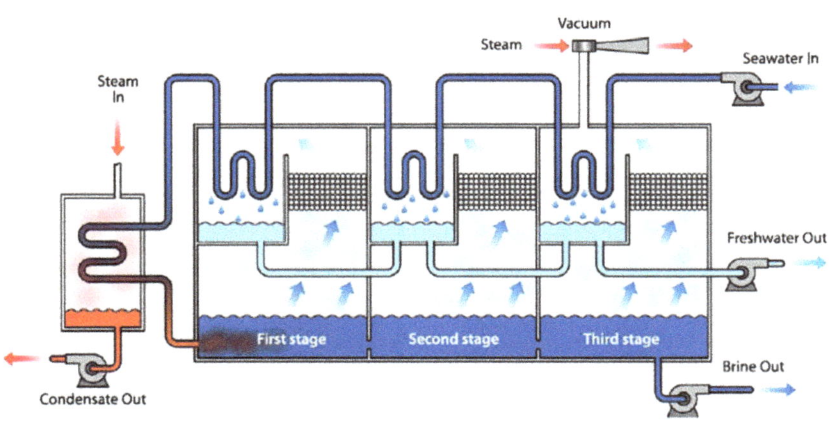

Figure 2.4. Schematic representation of the general MSF desalination process [16, 35]. Reprinted from [16], copyright (2021) with permission from Elsevier.

2.2.3 Vapour compressor distillation technique

The VCD method is often investigated as a mechanical vapour compression (MVC)/ recompression technology for desalinating water [34, 36]. Unlike other desalination methods, this type does not rely on external sources of heat; instead, it creates heat from the compressor [37, 41]. An external mechanical compressor is used to compress the free vapour produced when the saline water is heated to boiling in the evaporator. This is the core of the VCD's operating mechanism. Then, the compressed hot water vapour is routed back into the evaporator, condensing as freshwater outside the evaporator tubes. This process leaves behind hot brine water and the heat energy from the steam of distilled water, which is then used to boil the saline water [36]. The primary configurations of the water vapour compressor include the twin-screw version, which is discussed in [38, 39]; centrifugal [40, 41]; single-screw [42]; and roots compressor versions [43]. Several studies such as [44–46] were carried out to analyse the operational performance of the VCD approach by using the thermodynamic behaviour of the system. Thermo-economic study of single-effect and forward-feed multi-effect mechanical vapour compression desalination is reported in references [47, 48], respectively, whereas [49] investigated the performance of MVC driven by an organic Rankine cycle. An investigation of a system with zero emissions was carried out using MVC by [50]. The structure of VCD/MVC is depicted in figure 2.5 as a schematic diagram.

2.2.4 Multi-effect distillation

The design of MED plants incorporates several different airtight effects. At each effect, there is a progression of consistent evaporation and condensation that takes place in declining order of pressure and temperature. As the supply water goes

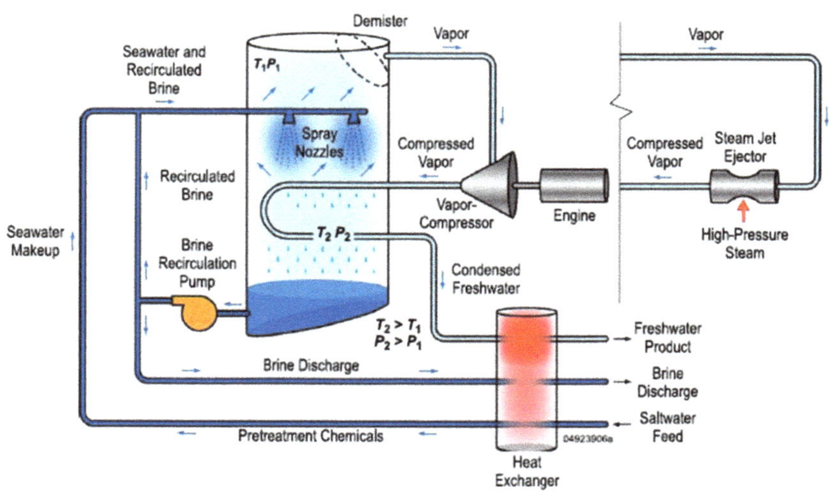

Figure 2.5. Schematic diagram of vapour compressor distillation/mechanical vapour compression [16, 35]. Reprinted from [16], copyright (2021) with permission from Elsevier.

through the first effect, which has the maximum pressure, the water is boiled by the heat provided from the outside. The vapour produced on the inside of the first effect is utilised as the energy source for the subsequent effect, and this process continues linearly along the series of effects. Therefore, on one side, the feed water is boiling and producing vapour, while on the other side, the vapour is condensing as freshwater [51]. Many different configurations of MED are based on an external heat source, hybridisation with other technologies, and a design arrangement of tubes, including vertical and horizontal tubes [33]. Figure 2.6 depicts a schematic diagram of MED. In [52], MED may be broken down into four configurations provided to the evaporators and the direction in which the vapour flows. Back-feed, parallel-feed, forward-feed, and parallel-feed with cross-feed (PCF) are the names given to these arrangements. The transient performance index of the MED system was utilised to make the comparisons between these setups. It was determined that PCF reacts more quickly than others do when external disturbances are introduced.

2.2.5 Electrodialysis and electrodialysis reversal methods

In electrolysis, both the positive and negative electrodes are placed in the same compartment; however, in ED, the positive and negative electrodes are placed in separate compartments. This contrasts with electrolysis, in which both electrodes are placed in the same compartment. Therefore, based on this information, the ED method of desalination means that direct current from the external source is delivered through the electrodes in a solution of salty water, including ion-selective membranes [53], and that these membranes are linked in parallel with one another to create channels. In the process of charging the electrodes, the negative salt ions flow through the anion permeable membrane in the direction of the anode, whilst the positive salt ions move in the opposite direction through the cation permeable membrane and make their way to the cathode. These motions of ions produce routes via which concentrated brine can escape, leaving behind only freshwater [33, 54]. The buildup of precipitant on the concentrated side is caused by the consistent usage of

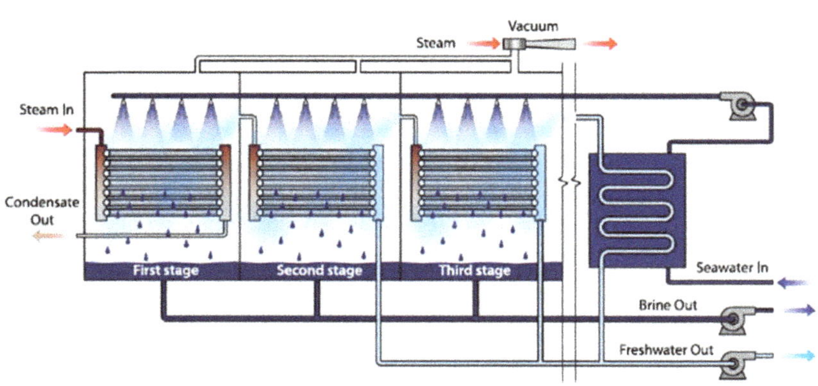

Figure 2.6. Schematic diagram of multi-effect distillation [16, 35]. Reprinted from [16], copyright (2021) with permission from Elsevier.

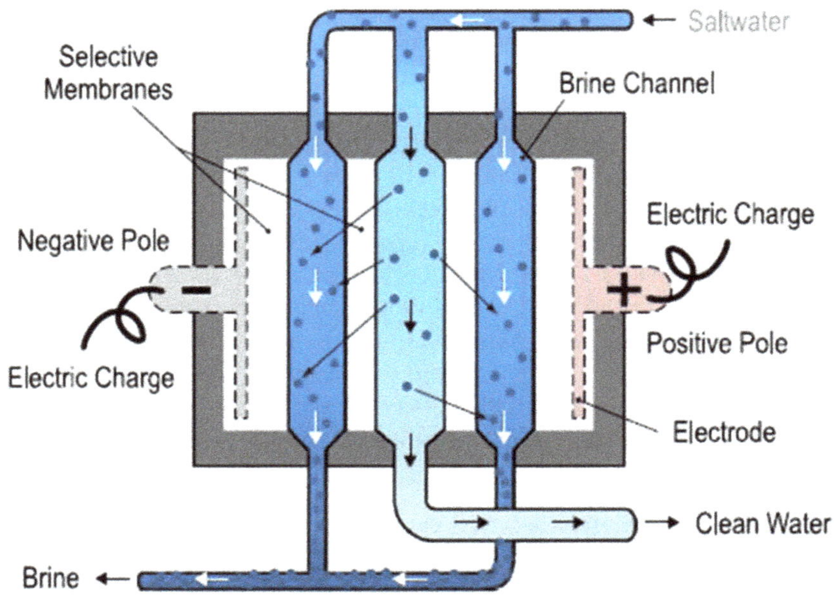

Figure 2.7. Schematic of electrodialysis desalination process [16, 35]. Reprinted from [16], copyright (2021) with permission from Elsevier.

membranes in the same direction over and over again. EDR offers a solution to this problem by regularly switching around the polarity of the electrodes. This reverses the flow of ions, which cleans out both sides of the membrane and extends its life span [33]. The schematic representation of the ED desalination process may be found in figure 2.7. Because of the drawbacks associated with electrodialysis (ED), such as the expensive cost of electrodes and membranes, the short life span of membranes, and the requirement for the recirculation of high-saline water in to acquire the necessary freshwater, there has been a reduced amount of interest and study conducted on this technique of desalination [55, 56]. The authors in [57] give a comprehensive analysis of the ED desalination technique, focusing on current advancements in the field and applications, one of which is the reverse ED method (RED). RED is employed in the process of extracting energy from differences in salinity. In this method, two solutions with differing salinity concentrations are mixed in a controlled manner, and the difference between them is then transformed into electricity. This has led to the creation of innovative designs and membranes and an improved method for plant layout suited for the operational processing of salty saltwater [58–60].

2.2.6 Capacitive deionisation technique

Both CDI and ED entail the removal of salt ions from salty water, resulting in freshwater as a byproduct of the process. In ED, an ion-selective membrane serves as the channel for removing salt ions, but carbon-coated electrodes serve in CDI as the channel for the adsorption of salt ions [61–63]. On the other hand, membrane capacitive deionisation (MCDI), which is an enhanced form of ED, uses an ion exchange membrane to separate positive and negative ions before attracting them to an electrode with the opposite polarity, which serves as the channel for adsorption [64–66]. The CDI process is designed so that salt ions will be adsorbed on the surface of the charged electrodes from the feed solution when an electrical potential is supplied between the electrodes from an external source. This will result in freshwater being left behind. At some point, the electrodes will become saturated, and the electric potentials will be reversed to regenerate the electrodes. This will result in the ions being desorbed back into the solution, producing brine [59]. The CDI can be constructed with either a unipolar or bipolar arrangement of electrodes, depending on the designer's preference, in the context of desalination [67–69].

In comparison to the requirements of other desalination processes, CDI requires very minimal upkeep. Using any chemicals or a pressure pump is unnecessary for this. Electrodes are more reliable and have a lower risk of being clogged than membranes. In addition to this, CDI has a high energy recovery rate, which might vary depending on the operating conditions [70–72]. Figure 2.8 depicts CDI and MCDI.

2.2.7 Membrane distillation technique

Desalination may be performed using MD, which is a hybrid approach that combines heat and membrane processes. It involves the use of a microporous hydrophobic membrane to accomplish the separation of volatile solutes from feed water. As a consequence of the temperature variations that occur on each side of the membrane, there is a difference in the vapour pressure of the component that is being separated [73]. The difference in vapour pressure and the temperature differential across the membrane are the driving force behind the separation. Condensation occurs on the cold side of the membrane surface, which is reached by vapour from the heated feed passing through the pores of the membrane and moving from the hot side to the cold side. Therefore, desalination is due in part to the balance of vapour and liquid in the membrane and in part to diffusion. One of the most notable aspects of MD is that the process is not affected by the salinity of the input water [72, 74, 75]. Furthermore, MD has low operating pressure and temperature [76, 77] and guarantees a purity level of 100% [74, 75]. The MD is divided into four different configurations based on the ways in which water vapour is collected on the side of the membrane used for condensation [75, 78]. Direct contact membrane distillation, air-gap membrane distillation, vacuum distillation, and sweep gas membrane distillation are the different membrane distillation processes. Figure 2.9 provides a representation of MD as a diagram.

2.2.8 Reverse osmosis technique

Osmosis operates on the fundamental premise that water will travel across a semipermeable membrane from an aqueous solution with a lower concentration to one with a higher concentration, therefore diluting the more concentrated solution until it reaches a concentration level that is in equilibrium. This method does not require a

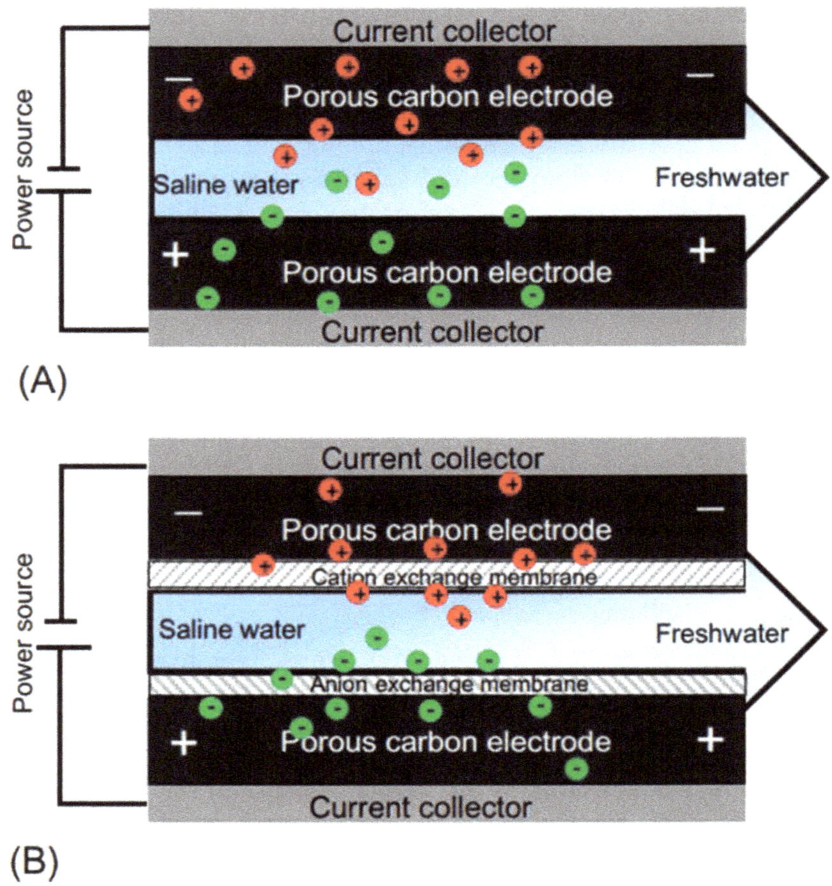

Figure 2.8. (A) Schematic diagram of capacitive deionisation and (B) Membrane capacitive membrane deionisation [16, 70]. Reprinted from [16], copyright (2021) with permission from Elsevier.

pressure difference greater than the osmotic pressure, which is a differential pressure threshold that applies to both the concentrated and the dilute solution. RO, on the other hand, as its name suggests, is the reverse process of osmosis in which water diffuses from the more concentrated to the less concentrated aqueous solution. This process is facilitated by a pressure differential across the semi-permeable membrane greater than the osmotic pressure [79]. For its part, the membrane is responsible for a significant pressure difference in relation to the flow of freshwater [80]. The large amount of energy required to maintain the pressurisation requirement of the RO technique is an obvious constraint; yet, RO is the method that is employed the most frequently because of its efficiency and overall cost-effectiveness [53, 81]. RO can also prevent practically all colloidal or solute particles from passing through the membrane; as a result, it guarantees that the freshwater produced is of high quality. According to [33], RO desalination consists of four different subsystems: (i)

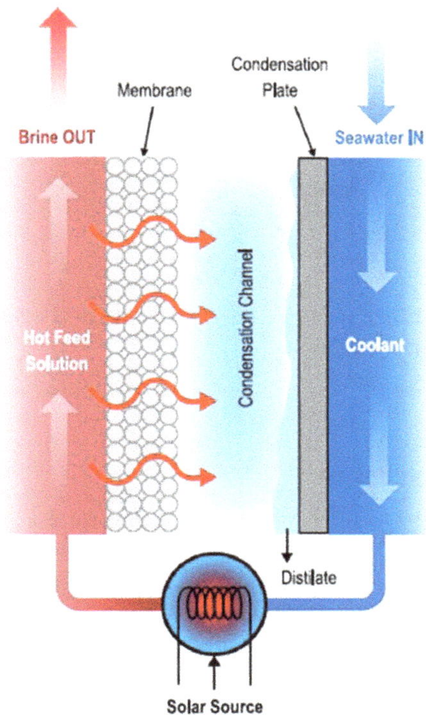

Figure 2.9. Schematic diagram of membrane distillation [16, 35]. Reprinted from [16], copyright (2021) with permission from Elsevier.

pretreatment section, which involves some form of percolation and sterilisation to reduce scaling and fouling by adding useful chemicals; (ii) a high pressure pump used for the creation of the required pressure differential necessary to push water across the membrane; (iii) a membrane for the separation of freshwater from concentrated feed water; and (iv) post-treatment with chemicals to guarantee high-quality output. The illustration of RO's structure is shown in figure 2.10.

The operational design of the RO system was defined and categorised into three primary configurations by [82]. When comparing these different design configurations, the link between SEC and the average water flow of the RO system was taken into account. On the basis of the average water flux and SEC, which were the primary criteria that were investigated, it was determined that a two-stage RO design has the potential to be regarded as the most suitable option. However, unlike the other two designs, this one requires significant financial investment.

There are several other desalination techniques, most of which are either an appendix to the primary processes or a mix of two methods. In certain circumstances, there are novel methods of desalination that have not yet undergone considerable research or been put into practise compared to the most common ones. FO, NF, freeze desalination (FD), and hybrid methods are all included in this category.

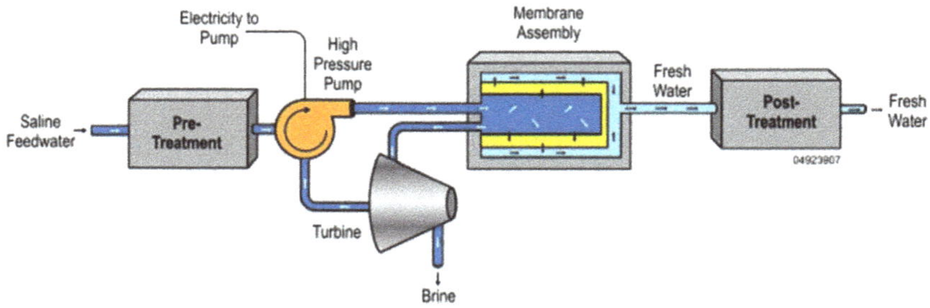

Figure 2.10. Schematic representation of the reverse osmosis desalination process [16, 35]. Reprinted from [16], copyright (2021) with permission from Elsevier.

2.2.9 Forward osmosis technique

It is possible to think of FO as the procedure diametrically opposed to RO. It applies the fundamental concept of osmosis, which depends on the natural osmotic pressure difference across the membrane. This causes water to be drawn from the side of the membrane containing a less concentrated solute (salty feed water) to the side containing a more concentrated solute (draw solute), thereby achieving a concentration equilibrium. According to [83], including a freshwater recovery unit is the only need for completing the desalination process when using FO as the feed stream. As a result, the ideal use for FO is as a pretreatment unit for use in conjunction with other desalination techniques such as RO. It has been found that the hybrid technique is superior to the solo methods [84]. RO is not a suitable method for the desalination of high-salinity input water; however, the hybrid FO desalination system is a suitable method. The energy requirements of the hybrid FO system are lower than the RO systems. High water recovery and minimal fouling are two additional benefits of using FO instead of RO [85, 86]. When used as a solo desalination technique, FO has significant drawbacks, the most significant of which are high energy consumption and a restricted selection of draw solutes [87].

2.2.10 Nanofiltration

The removal of a scaling divalent ion requires that the separation process be based on the solute charge and the molecular size of the uncharged solute. NF is a desalination technique that uses a membrane, and the pore diameters of the membrane are less than 0.002 micrometres [88]. In contrast to RO, NF membranes can selectively reject solutes depending on their charge and size. This approach also needs a lower pressure for normal operation than RO does. It is not sufficient for the desalination of saltwater; rather, it should be used as a pretreatment unit in a hybrid technique [89]. NF can be configured in two different ways, depending on the flow pattern that occurs across the membrane: (i) crossflow with concentrate recycling, also known as tangential flow or dynamic filtration; and (ii) a dead-end flow system, also known as static filtration or dead-end filtration [85]. Crossflow with concentrate recycle is the most common of the two configurations. The creation and enhancement of nanomaterials are gradually becoming the main focus of research trends in this field [90].

2.2.11 Freezing desalination

The FD technique operates on the principle of phase change, mainly the transition from liquid to solid (also known as solidification). Feed water, which is saline water, goes through a freezing process, during which a refrigerant is used to make ice crystals mostly made of freshwater. This process leaves behind brine, which is a concentrated liquid. Following this step, the ice slurries are extracted from the brine before being melted to produce the end product, which is freshwater [91]. Therefore, the FD process consists of three phases, which are the precooling, crystallisation, and separation steps [88, 92]. There are two different types of FD setups, and the location of the refrigerant distinguishes each one. Both direct and indirect FDs are included in this number [88]. When the names suggest, the direct FD design places the refrigerant in direct contact with the feed water. As a result, the refrigerant can absorb heat from the feed water when the water is vaporised. On the other hand, indirect FD utilises a crystalliser wall that acts as a barrier between the feed water and the refrigerant. Although direct FD is believed to be more effective than indirect FD due to its wide surface area, high heat transfer coefficient, and economic feasibility, its drawbacks include the possibility of hydrate formation and water contamination by refrigerant [88].

2.3 Summary

This chapter reviews major water treatment techniques, highlighting their design configurations, processes, merits, and drawbacks. Wastewater and reuse techniques were briefly discussed as the primary and secondary methods are mainly straightforward processes, and the advanced methods are like those of desalination processes. Therefore, desalination utilises more advanced water treatment processes, categorised into thermal and membrane processes. The thermal process involves methods of desalination that require a change of state and are dependent on both electrical and heat energy to achieve this. However, the membrane process comprises methods of desalination that use permeable membranes to separate freshwater from saline water. These methods do not require a change of state and therefore depend mainly on electrical energy that can create a pressure difference across the membrane. As a result, the membrane process requires less energy than the thermal process of desalination. This makes methods such as RO very popular. The RO method is the most widely implemented desalination method because of its low energy requirement and simple configuration that allows easy hybridisation with other methods. Hybridisation of different methods proves to be more effective as the disadvantage of one method may be supplemented by the merit of another [93]. The various approaches to desalination are broken down and compared in tables 2.1 and 2.2.

Table 2.1. Summary of the comparison of desalination methods. Reprinted from [16], copyright (2021).

Desalination processes	Methods of desalination	Energy consumption	Merits	Demerits	Ref.
Thermal processes	HDH		Low-grade thermal energy may be used and the basic design means that it only has to make use of inexpensive materials.	There are difficulties in scaling up the energy demand because it is greater than that of the MED, MSF, and membrane techniques.	[21, 22]
	MSF	19.58–27.25 kWh m^{-3}	When dealing with high concentrations of total dissolved solids (TDS), this method proves to be far more effective than reverse osmosis. Simple to operate and manage.	Low water recovery compared to RO and higher energy demand compared to membrane and MED. The typical operation rate is a bit above 60%. Hot brine is released into the environment.	[94, 95, 96]
	VCD	7–12 kWh m^{-3} for MVC and 16.26 kWh m^{-3} for thermal vapour compressor	The compressor provides the necessary heat. Energy consumption is lower than that of MFS and MED.	The VCD capacity is relatively modest, making it less cost-effective than MFS and MED. Therefore, it is recommended to combine this approach with others.	[36, 91]
	FD		High water flow values, permeate flux, and water recovery rate; low operating pressure; low maintenance cost.	The likelihood that the refrigerant may form hydrates and contaminate water supplies.	[88, 89]
	MED	14.45–21.35 kWh m^{-3}	This method is superior to RO in cases of high total dissolved solids. Efficient in reusing waste heat.	Hot brine is released into the environment and energy consumption is higher than with membrane techniques.	[91, 92, 17]
Membrane processes	ED	2.64–5.5, 0.7–2.5 kWh m^{-3} at low TDS	Compared to thermal processes, ED requires less energy and costs less to implement. Unlike RO, which requires external pressure, electrical potential deference. RED allows for greater water recovery and decreased membrane fouling and scaling.	Practical for brackish water. Electrodes corrode rapidly, leading to high costs and a limited life span. Reusing highly saline water to supplement freshwater supplies.	[51, 52, 56]

		techniques, it has low maintenance needs and does not require any special chemicals or pumps. Depending on the operating conditions, it can recover a lot of energy quickly because of its stable electrodes and low rate of fouling.	expensive; limited electrode life owing to corrosion.	67, 70]
MD		Total dissolved solids in input water as high as 200 000 mg l^{-1} can be effectively removed. It features a straightforward plant layout and less membrane fouling than RO. It has a high rejection rate and recovers energy from rejected heat.	Lack of specialised membrane; high electrical and thermal energy demand; low water recovery and water flow with a likelihood of membrane wetness.	[71, 72, 76]
RO	4–6 kWh m^{-3} for SWRO (seawater reverse osmosis) and 1.5–2.5 kWh m^{-3} for BWRO (brackish water reverse osmosis)	It requires less energy, produces less waste heat, recovers more water, costs less to run, and is less expensive to build. In comparison to other membrane techniques, RO's membrane costs less.	Membrane scaling and fouling due to high external pressure; brine's chemicals are used, which might have a harmful effect on the environment.	[77, 92, 17]
FO		The oscillating pressure differential and flow resistance within the membrane module are all that is needed to drive the process, meaning that no external pressure is necessary. Strong resistance to a wide range of pollutants.	High expense due to freshwater extraction and draw solution regeneration and control system to avoid cross-contamination between feed and draw solution. Lower flow rate compared to reverse osmosis. It requires a concentrated draw solution, which requires easy separation.	[84, 82]
NF	3.35 kWh m^{-3}	It operates at a pressure that is less than that needed for RO. The permeate flow rate and feed volume are high, and water recovery is excellent. Feed does not need to be warmed, cooled, or stirred mechanically. In a hybrid setup, it is most useful as a pretreatment unit.	Unlike RO, which rejects all solutes, NF selectively rejects solutes depending on charge and size. Reduced longevity of the membrane. It is not powerful enough to desalinate ocean water.	[85, 87]

Table 2.2. Hybridisation of RO and other methods of desalination. Reprinted from [16], copyright (2021).

Hybrid methods	Performance indicators	Limitations	References
HDH-RO	Enhance the temperature of saline recirculation to be higher than the system's make-up temperature. Reduction in energy consumption by HDH alone.	17.5% higher in the total investment cost of the RO unit.	[97, 98, 99]
MSF-RO	Use as cogeneration power desalination plants to produce both electricity and freshwater.	Complex to operate.	[30, 100 101, 102]
TVC-RO	Improvement of SEC and production ratio.	Possible limited distribution ratio due to insufficient input water from the thermal vapour compressor (TVC) to the RO unit.	[43]
MED-RO	Production of high-quality water, high recovery ratio, minimal energy consumption, and RO saline treatment.	Dependent on the steam's quality. Scale formation on the surface of a heat exchanger.	[103, 104, 105]
RO-ED	Minimal brine production and utilisation for salt production.	The focus tends to shift towards salt production.	[106, 107].
RO-CDI	High-quality water production.	Delay in the charge and discharge of the CDI cell.	[108]
RO-MD	Use for treating brine.	Fouling and membrane moisture are significant issues that reduce flux.	[102, 109, 110, 111]
FO-RO	Use for the treatment of brine and wastewater for reuse.	Recovery of the draw solute after FO treatment of RO brine is difficult.	[102, 112]
RO-NF (nanofiltration)	Improved water quality while using reduced RO pressure and energy consumption.	Increased complexity and cost with the use of more membrane.	[77, 113]
RO-FD	Brine treatment and management.	Nucleation control within the FD unit.	[88, 114]

References

[1] Tortajada C 2020 Contributions of recycled wastewater to clean water and sanitation sustainable development goals *NBJ Clean Water* **3** 22

[2] European Union Regional Development FundAQUARESRRAPK 2020 Water reuse policies advancement for resource efficient AQUARES A1.3 – Water reuse technology application guide 2020 *Water Reuse Technology Application Guide*

[3] Irfan Z B 2016 Urban water reuse risk *Urban Water Reuse Handbook* ed S Eslamian (Boca Raton, FL: CRC Press)
[4] Henkanatte-Gedera S M, Selvaratnam T, Caskan N, Nirmalakhandan N, Van Voorhies W and Lammers P J 2015 Algal-based, single-step treatment of urban wastewaters *Bioresour. Technol.* **189** 273–8
[5] La Manna P, De Carluccio M, Iannece P, Vigliotta G, Proto A and Rizzo L 2023 Chelating agents supported solar photo-Fenton and sunlight/H_2O_2 processes for the removal of pharmaceuticals and resistant pathogens inactivation in quaternary treatment for the reuse of urban wastewater reuse *J. Hazard. Mater.* **452** 131235
[6] Selvaratnam T *et al* 2014 Evaluation of a thermo-tolerant acidophilic alga, Galdieria sulphuraria, for nutrient removal from urban wastewaters *Bioresour. Technol.* **156** 395–9
[7] Lugo A *et al* 2023 Life cycle energy use and greenhouse gas emissions for a novel algal-osmosis membrane system versus conventional advanced potable water reuse processes: part I *J. Environ. Manage.* **331** 117293
[8] Cosenza A, Gulhan H, Maida C M and Mannina G 2022 Nutrient recovery from wastewater treatment by ultrafiltration membrane for water reuse in view of a circular economy perspective *Bioresour. Technol.* **363** 127929
[9] Delanka-Pedige H M K *et al* 2019 Pathogen reduction in an algal-based wastewater treatment system employing Galdieria sulphuraria *Algal Res* **39** 101423
[10] Jimoh T A, Laubscher R K, Askew D J and Cowan A K 2021 Advanced oxidation as tertiary treatment for recovery of effluent from an integrated algal pond system *Water Environ. J.* **35** 1094–102
[11] Xu P, Janex M L, Savoye P, Cockx A and Lazarova V 2002 Wastewater disinfection by ozone: main parameters for process design *Water Res.* **36** 1043–55
[12] Pérez G, Gómez P, Ortiz I and Urtiaga A 2022 Techno-economic assessment of a membrane-based wastewater reclamation process *Desalination* **522** 115409
[13] Stefanakis A I 2015 Modern water reuse technologists: tertiary membrane and activated carbon filtration *Urban Water Reuse Handbook* ed S Eslamian (Boca Raton, FL: CRC Press) pp 419–28
[14] Abdelkareem M A, El Haj Assad M, Sayed E T and Soudan B 2018 Recent progress in the use of renewable energy sources to power water desalination plants *Desalination* **435** 97–113
[15] Okampo E J and Nwulu N I 2020 Optimal energy mix for a reverse osmosis desalination unit considering demand response osmosis *J. Eng. Des. Technol.* **18** 1287–303
[16] Okampo E J and Nwulu N 2021 Optimisation of renewable energy powered reverse osmosis desalination systems: a state-of-the-art review *Renew. Sustain. Energy Rev.* **140** 110712
[17] Ahmed F E, Hashaikeh R and Hilal N 2019 Solar powered desalination—technology, energy and future outlook *Desalination* **453** 54–76
[18] Yildirim C and Solmuş I 2014 A parametric study on a humidification-dehumidification (HDH) desalination unit powered by solar air and water heaters *Energy Convers. Manag.* **86** 568–75
[19] Ettouney H 2005 Design and analysis of humidification dehumidification desalination process *Desalination* **183** 341–52
[20] Capocelli M, Balsamo M, Lancia A and Barba D 2018 Process analysis of a novel humidification-dehumidification-adsorption (HDHA) desalination method *Desalination* **429** 155–66

[21] Narayan G P, Summers E K, Lienhard J H, Sharqawy M H, Antar M A and Zubair S M 2009 The potential of solar-driven humidification–dehumidification desalination for small-scale decentralised water production *Renew. Sustain. Energy Rev.* **14** 1187–201
[22] Segurado R, Costa M, Duić N and Carvalho M G 2014 Integrated analysis of energy and water supply in islands. Case study of S. Vicente, Cape Verde *Energy* **92** 639–48
[23] Müller-Holst H, Engelhardt M and Schölkopf W 1999 Small-scale thermal seawater desalination simulation and optimisation of system design *Desalination* **122** 255–62
[24] Müller-Holst H 2007 Solar thermal desalination using the multiple effect humidification (MEH)-method *Solar Desalination for the 21st Century* ed L Rizutti, H M Ettouney and A Cipollina (Berlin: Springer) pp 215–25
[25] Lawal D, Antar M, Khalifa A, Zubair S and Al-Sulaiman F 2018 Humidification-dehumidification desalination system operated by a heat pump *Energy Convers. Manag.* **161** 128–40
[26] Bandi C S, Uppaluri R and Kumar A 2016 Global optimization of MSF seawater desalination processes *Desalination* **394** 30–43
[27] Darawsheh I, Islam M D and Banat F 2019 Experimental characterization of a solar powered MSF desalination process performance *Therm. Sci. Eng. Prog.* **10** 154–62
[28] Hawaidi E A M and Mujtaba I M 2010 Simulation and optimization of MSF desalination process for fixed freshwater demand: impact of brine heater fouling *Chem. Eng. J.* **165** 545–53
[29] El-Dessouky H, Alatiqi I and Ettouney H 1998 Process synthesis: the multi-stage flash desalination system *Desalination* **115** 155–79
[30] Tanvir M S and Mujtaba I M 2008 Optimisation of design and operation of MSF desalination process using MINLP technique in gPROMS *Desalination* **222** 419–30
[31] Abduljawad M and Ezzeghni U 2010 Optimization of Tajoura MSF desalination plant *Desalination* **254** 23–8
[32] Marcovecchio M G, Mussati S F, Aguirre P A and Scenna N J 2005 Optimization of hybrid desalination processes including multi stage flash and reverse osmosis systems *Desalination* **182** 111–22
[33] Deyab M A 2019 Enhancement of corrosion resistance in MSF desalination plants during acid cleaning operation by cationic surfactant *Desalination* **456** 32–7
[34] Mannan M, Alhaj M, Mabrouk A N and Al-Ghamdi S G 2019 Examining the life-cycle environmental impacts of desalination: a case study in the State of Qatar *Desalination* **452** 238–46
[35] Al-Karaghouli A and Kazmerski L L 2013 Energy consumption and water production cost of conventional and renewable-energy-powered desalination processes *Renew. Sustain. Energy Rev.* **24** 343–56
[36] Shen J, Feng G, Xing Z and Wang X 2019 Theoretical study of two-stage water vapour compression systems *Appl. Therm. Eng.* **147** 972–82
[37] Saidur R, Elcevvadi E T, Mekhilef S, Safari A and Mohammed H A 2011 An overview of different distillation methods for small scale applications *Renew. Sustain. Energy Rev.* **15** 4756–64
[38] Xing Z, Zhang K, Wang X, He Z and Shen J 2015 Development of a water-injected twin-screw compressor for mechanical vapour compression desalination systems *Appl. Therm. Eng.* **95** 125–35
[39] Wang C, Shen J, Wang X, Xing Z and Tian Y 2017 Modeling and performance study of a water-injected twin-screw water vapour compressor *Int. J. Refrig.* **83** 75–87

[40] Veza J 1995 Mechanical vapour compression desalination plants—a case study *Desalination* **101** 1–10
[41] Kronenberg G and Lokiec F 2001 Low-temperature distillation processes in single-and dual-purpose plants *Desalination* **136** 189–97
[42] Yang J, Zhang C, Zhang Z, Yang L and Lin W 2016 Study on mechanical vapor recompression system with single screw wet compression single screw compressor *Appl. Therm. Eng.* **103** 205–11
[43] Bahar R, Hawlader M N A and Woei L S 2004 Performance evaluation of a mechanical vapour compression desalination system *Desalination* **166** 123–7
[44] Chen Q, Ja M K, Li Y and Chua K J 2019 Energy, exergy and economic analysis of a hybrid spray-assisted low-temperature desalination/thermal vapour compression system *Energy* **166** 871–85
[45] Cao Z, Deng J, Ye F and Garris C A 2018 Analysis of a hybrid thermal vapor compression and reverse osmosis desalination system at variable design conditions *Desalination* **438** 54–62
[46] zhuo Jin C, li Chou Q, sheng Jiao D and cheng Shu P 2014 Vapour compression flash seawater desalination system and its exergy analysis *Desalination* **353** 75–83
[47] Jamil M A and Zubair S M 2017 On thermoeconomic analysis of a single-effect mechanical vapor compression desalination system *Desalination* **420** 292–307
[48] Jamil M A and Zubair S M 2017 Design and analysis of a forward feed multi-effect mechanical vapour compression desalination system: an exergo-economic approach *Energy* **140** 1107–20
[49] He W F, Ji C, Han D, Wu Y K, Huang L and Zhang X K 2017 Performance analysis of the mechanical vapour compression desalination system driven by an organic Rankine cycle *Energy* **141** 1177–86
[50] Han D, He W F, Yue C and Pu W H 2017 Study on desalination of zero-emission system based on mechanical vapour compression *Appl. Energy* **185** 1490–6
[51] Garg M C 2019 Renewable energy-powered membrane technology: cost analysis and energy consumption *Current Trends and Future Developments in (Bio-) Membranes: Renewable Energy Integrated with Membrane Operations* ed A Basile, A Cassano and A Figoli (New York: Elsevier) pp 85–110
[52] Elsayed M L, Mesalhy O, Mohammed R H and Chow L C 2018 Transient performance of MED processes with different feed configurations *Desalination* **438** 37–53
[53] Okampo E J and Nwulu N 2020 Optimal design and techno-economic evaluation of renewable energy powered combined reverse osmosis desalination and brine treatment unit *Desalin. Water Treat.* **202** 27–37
[54] Al-Karaghouli A, Renne D and Kazmerski L L 2010 Technical and economic assessment of photovoltaic-driven desalination systems *Renew. Energy* **35** 323–8
[55] Lee H J, Sarfert F, Strathmann H and Moon S H 2002 Designing of an electrodialysis desalination plant *Desalination* **142** 267–86
[56] Ishimaru N 1994 Solar photovoltaic desalination of brackish water in remote areas by electrodialysis *Desalination* **98** 485–93
[57] Campione A, Gurreri L, Ciofalo M, Micale G, Tamburini A and Cipollina A 2018 Electrodialysis for water desalination: a critical assessment of recent developments on process fundamentals, models and applications *Desalination* **434** 121–60
[58] Daniilidis A, Vermaas D A, Herber R and Nijmeijer K 2014 Experimentally obtainable energy from mixing river water, seawater or brines with reverse electrodialysis *Renew. Energy* **64** 123–31

[59] Cerva M L L *et al* 2017 Coupling CFD with a one-dimensional model to predict the performance of reverse electrodialysis stacks *J. Memb. Sci.* **541** 595–610

[60] Tedesco M, Cipollina A, Tamburini A and Micale G 2017 Towards 1 kW power production in a reverse electrodialysis pilot plant with saline waters and concentrated brines *J. Memb. Sci.* **522** 226–36

[61] Kress N 2019 Desalination technologies *Marine Impacts of Seawater Desalination: Science, Management, and Policy* (New York: Elsevier) pp 11–34

[62] Ahualli S, Iglesias G R and Delgado V 2018 Principles and theoretical models of CDI: experimental approaches *Interface Sci. Technol.* **24** 169–92

[63] Tang K *et al* 2019 Seawater desalination by over-potential membrane capacitive deionization: opportunities and hurdles *Chem. Eng. J.* **357** 103–11

[64] Saleem M W and Kim W S 2018 Parameter-based performance evaluation and optimisation of a capacitive deionization desalination process *Desalination* **437** 133–43

[65] Lee J B, Park K K, Eum H M and Lee C W 2006 Desalination of a thermal power plant wastewater by membrane capacitive deionization *Desalination* **196** 125–34

[66] Wang Z, Gong H, Zhang Y, Liang P and Wang K 2017 Nitrogen recovery from low-strength wastewater by combined membrane capacitive deionisation (MCDI) and ion exchange (IE) process *Chem. Eng. J.* **316** 1–6

[67] Chen Y A, Fan C S and Hou C H 2018 Optimizing the energetic performance of capacitive deionisation devices with unipolar and bipolar connections under constant current charging *J. Taiwan Inst. Chem. Eng.* **93** 201–10

[68] Liu X *et al* 2016 Bipolarly stacked electrolyser for energy and space efficient fabrication of supercapacitor electrodes *J. Power Sources* **307** 208–13

[69] Zhang C, He D, Ma J, Tang W and Waite T D 2018 Faradaic reactions in capacitive deionisation (CDI)—problems and possibilities: a review *Water Res.* **128** 314–30

[70] Sen Gupta S, Islam M R and Pradeep T 2019 Capacitive deionisation (CDI): an alternative cost-effective desalination technique *Advances in Water Purification Techniques: Meeting the Needs of Developed and Developing Countries* ed S Ahuja (New York: Elsevier) pp 165–202

[71] Kang J, Kim T, Shin H, Lee J, Ha J I and Yoon J 2016 Direct energy recovery system for membrane capacitive deionisation *Desalination* **398** 144–50

[72] Zhao R, Porada S, Biesheuvel P M and Van der Wal A 2013 Energy consumption in membrane capacitive deionisation for different water recoveries and flow rates, and comparison with reverse osmosis *Desalination* **330** 35–41

[73] Soni V, Abildskov J, Jonsson G and Gani R 2009 A general model for membrane-based separation processes *Comput. Chem. Eng.* **33** 644–59

[74] González D, Amigo J and Suárez F 2017 Membrane distillation: perspectives for sustainable and improved desalination *Renew. Sustain. Energy Rev.* **80** 238–59

[75] Al-Obaidani S, Curcio E, Macedonio F, Di Profio G, Al-Hinai H and Drioli E 2008 Potential of membrane distillation in seawater desalination: thermal efficiency, sensitivity study, and cost estimation *J. Memb. Sci.* **323** 85–98

[76] Winter D, Koschikowski J and Wieghaus M 2011 Desalination using membrane distillation: experimental studies on full-scale spiral wound modules *J. Memb. Sci.* **375** 104–12

[77] Ashoor B B, Mansour S, Giwa A, Dufour V and Hasan S W 2016 Principles and applications of direct contact membrane distillation (DCMD): a comprehensive review *Desalination* **398** 222–46

[78] Goh P S, Matsuura T, Ismail A F and Hilal N 2016 Recent trends in membranes and membrane processes for desalination *Desalination* **391** 43–60
[79] Fritzmann C, Lowenberg J, Wintgens T and Melin T 2007 State o the art of reverse osmosis desalination *Desalination* **216** 1–76
[80] Abdallah S, Abu-Hilal M and Mohsen M S 2005 Performance of a photovoltaic powered reverse osmosis system under local climatic conditions *Desalination* **183** 95–104
[81] Okampo E J and Nwulu N 2021 Techno-economic evaluation of reverse osmosis desalination system considering emission cost and demand response *Sustain. Energy Technol. Assess.* **46** 101252
[82] Lin S and Elimelech M 2017 Kinetics and energetics trade-off in reverse osmosis desalination with different configurations *Desalination* **401** 42–52
[83] Jaafar J *et al* 2018 Forward osmosis for desalination application *Membrane Separation Principles and Applications* (Handbooks in Separation Science) ed A F Ismail, M A Rahman, M H D Othman and T Matsuura (New York: Elsevier) pp 315–37
[84] Syed Ibrahim G P, Isloor A M and Yuliwati E 2019 *Current Trends and Future Developments on (Bio-) Membranes* ed A Basile, E Curcio and Inamuddin (New York: Elsevier)
[85] Peñate B and García-Rodríguez L 2012 Current trends and future prospects in the design of seawater reverse osmosis desalination technology *Desalination* **284** 1–8
[86] Choi Y J, Choi J S, Lee S, Yang D R and Kim J H 2009 Toward a combined system of forward osmosis and reverse osmosis for seawater desalination *Desalination* **247** 239–46
[87] Akther N, Sodiq A, Giwa A, Daer S, Arafat H A and Hasan S W 2015 Recent advancements in forward osmosis desalination: a review *Chem. Eng. J.* **281** 502–422
[88] Abdel-Fatah M A 2018 Nanofiltration systems and applications in wastewater treatment: review article *Ain Shams Eng. J.* **9** 3077–92
[89] Matin A, Rahman F, Shafi H Z and Zubair S M 2019 Scaling of reverse osmosis membranes used in water desalination: phenomena, impact, and control; future directions *Desalination* **455** 135–57
[90] Teow Y H and Mohammad A W 2019 New generation nanomaterials for water desalination: a review *Desalination* **451** 2–17
[91] Kalista B, Shin H, Cho J and Jang A 2018 Current development and future prospect review of freeze desalination *Desalination* **447** 167–81
[92] Cao W, Beggs C and Mujtaba I M 2014 Theoretical approach to the desalination of freeze seawater desalination on flake ice maker utilizing LNG cold energy *Desalination* **355** 22–32
[93] Okampo E J and Nwulu N 2022 Economic and reliability assessment of hybrid PRO-RO desalination systems using brine for salinity gradient energy production *Sustainability* **14** 3328
[94] Mezher T, Fath H, Abbas Z and Khaled A 2011 Techno-economic assessment and environmental impacts of desalination technologies *Desalination* **266** 263–73
[95] Shahzad M W, Burhan M, Ang L and Ng K C 2017 Energy-water-environment nexus underpinning future desalination sustainability *Desalination* **413** 52–64
[96] Mayor B 2019 Growth patterns in mature desalination technologies and analogies with the energy field *Desalination* **457** 75–84
[97] Jamil M A, Elmutasim S M and Zubair S M 2018 Exergo-economic analysis of a hybrid humidification dehumidification reverse osmosis (HDH-RO) system operating under different retrofits *Energy Convers. Manag.* **158** 286–97

[98] Narayan G P, McGovern R K, Zubair,Zubair S M and Lienhard J H 2012 High-temperature-steam-driven, varied-pressure, humidification-dehumidification system coupled with reverse osmosis for energy-efficient seawater desalination *Energy* **37** 482–93
[99] Eslamimanesh A and Hatamipour M S 2010 Economical study of a small-scale direct contact humidification-dehumidification desalination plant *Desalination* **250** 203–7
[100] Almulla A, Hamad A and Gadalla M 2005 Integrating hybrid systems with existing thermal desalination plants *Desalination* **174** 171–92
[101] Bandi C S, Uppaluri R and Kumar A 2016 Global optimality of hybrid MSF-RO seawater desalination processes *Desalination* **400** 47–59
[102] Heidary B, Hashjin T T, Ghobadian B and Roshandel R 2018 Optimal integration of small-scale hybrid solar wind RO-MSF desalination system *Renew. Energy Focus* **27** 120–34
[103] Filippini G, Al-Obaidi M A, Manenti F and Mujtaba I M 2018 Performance analysis of hybrid system of multi effect distillation and reverse osmosis for seawater desalination via modelling and simulation *Desalination* **448** 21–35
[104] Khoshgoftar Manesh M H, Ghalami H, Amidpour M and Hamedi M H 2013 Optimal coupling of site utility steam network with MED-RO desalination through total site analysis and exergoeconomic optimisation *Desalination* **316** 42–52
[105] Lee S, Choi J, Park Y G, Shon H, Ahn C H and Kim S H 2019 Hybrid desalination processes for beneficial use of reverse osmosis brine: current status and future prospects *Desalination* **454** 104–11
[106] Nayar K G, Fernandes J, McGovern R K, Al-Anzi B S and Lienhard J H 2019 Cost and energy needs of RO-ED-crystallizer systems for zero brine discharge seawater desalination *Desalination* **457** 115–32
[107] Thiel G P, Kumar A, Gómez-González A and Lienhard J H 2017 Utilisation of desalination brine for sodium hydroxide production: technologies, engineering principles, recovery limits, and future directions *ACS Sustain. Chem. Eng.* **5** 11147–62
[108] Minhas M B, Jande Y A C and Kim W S 2014 Combined reverse osmosis and constant-current operated capacitive deionisation system for seawater desalination *Desalination* **344** 299–305
[109] Choi Y *et al* 2017 Experimental comparison of submerged membrane distillation configurations for concentrated brine treatment *Desalination* **420** 54–62
[110] Yan Z *et al* 2017 Reverse osmosis brine treatment using direct contact membrane distillation: effects of feed temperature and velocity *Desalination* **423** 149–56
[111] Sanmartino J A, Khayet M, García-Payo M C, El-Bakouri H and Riaza A 2017 Treatment of reverse osmosis brine by direct contact membrane distillation: chemical pretreatment approach *Desalination* **420** 79–90
[112] Ge J, Peng Y, Li Z, Chen P and Wang S 2014 Membrane fouling and wetting in a DCMD process for RO brine concentration *Desalination* **344** 97–107
[113] Zhou D, Zhu L, Fu Y, Zhu M and Xue L 2015 Development of lower cost seawater desalination processes using nanofiltration technologies—a review *Desalination* **376** 109–16
[114] Randall D G and Nathoo J 2015 A succinct review of the treatment of reverse osmosis brines using freeze crystallization *J. Water Process Eng.* **8** 186–94

Part III

Optimisation of the desalination system with renewable energy sources

IOP Publishing

Ewaoche John Okampo, Nnamdi Nwulu and Pitshou Ntambu Bokoro

Chapter 3

Potential of renewable energy sources for desalination and water reuse

As the demand for freshwater continues to escalate in the face of growing population, urbanisation, and climate change impacts, the intersection of water scarcity and energy sustainability has never been more critical. Traditional water treatment and desalination processes are energy-intensive, often relying on fossil fuels and contributing to greenhouse gas emissions. However, the integration of renewable energy sources with desalination and water reuse technology holds immense promise in addressing these challenges while fostering a more sustainable water-energy nexus. This chapter explores different renewable energy sources (RES) available and suitable for desalination system. The availability of RES depends on the location while its suitability depends on the desalination process. Solar energy is identified to be the most used of RES for desalination but its irradiation leaves room for uncertainty and unreliability. Thus, hybrid energy sources are more sustainable for desalination systems.

3.1 Introduction

The high energy demand of the desalination thermal process and the subsequent negative impact on the environment caused by the conventional energy sources that are utilised to power the thermal techniques have caused the membrane process powered by RES to become increasingly dominant and common. However, because there are so many different types of RES and so many different ways to desalinate water, it is not a simple task to decide which type of renewable energy source will work best with a particular desalination technique. According to [1], the selection of RES for desalination depends on several factors, including the size of the desalination plant, its location, the feed pressure and characteristics, and the anticipated cost of producing freshwater. The intermittent and low-intensity output that is typical of

doi:10.1088/978-0-7503-6154-5ch3

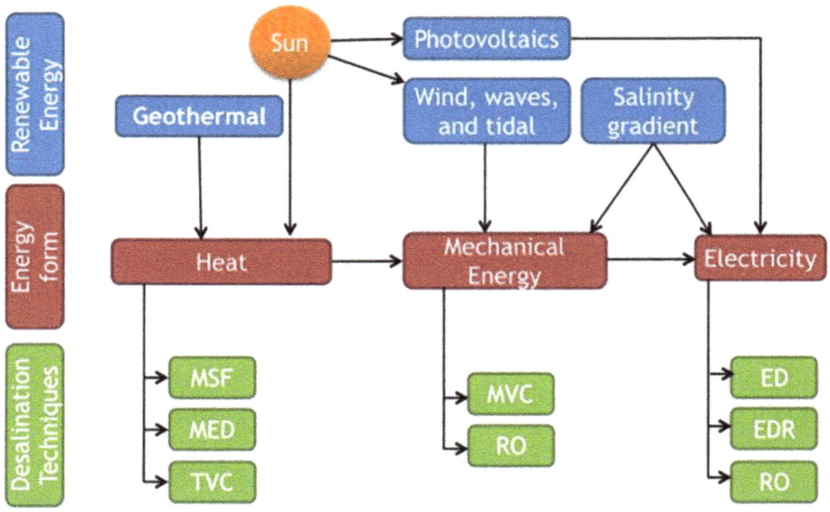

Figure 3.1. Possible integration of RES with desalination methods [1, 7]. Abbreviations: MED, multi-effect distillation; MVC, mechanical vapour compression; RO, reverse osmosis; MSF, multi-stage flash distillation; TVC, thermal vapour compression; ED, electrodialysis; EDR, electrodialysis reversal. Reprinted from [7], copyright (2021) with permission from Elsevier.

many forms of RES is one of the factors that makes their utilisation problematic. Integration with the grid, hybridisation, and the utilisation of energy storage devices or batteries are three approaches that can reduce the severity of these difficulties or possibly eliminate them altogether. The potential integration of RES with desalination technologies is illustrated in figure 3.1, which is based on the reference content [1].

In the next part, we will talk about a few of the RES that are now available and are being investigated for use in the desalination system.

3.2 Geothermal energy potential for desalination

Geothermal energy is the source of heat from below the earth's surface. This heat can take the form of hot water or it can steam, or be a combination of the two. This kind of renewable energy source is divided into two categories according to the temperature range, or the enthalpy range: (i) low temperature, which refers to a temperature range that is less than 150 °C, and (ii) high temperature, which refers to a temperature range that is greater than 150 °C [2]. It is one of the RES that is the most reliable and least harmful to the environment. It delivers a constant heat flow, which contributes to its dependability. It is not affected by changes in weather, and the costs associated with running it are very reasonable. These characteristics demonstrate the value of geothermal energy and make it an excellent choice for the desalination process.

The primary obstacle is the expensive construction of new power plants. In addition, its activity can occur only in a restricted number of locations [1].

Geothermal energy has a greater potential for use in thermal desalination procedures due to the ability of geothermal facilities to directly deliver heat energy. In Tunisia, humidification-dehumidification and membrane distillation were powered by geothermal energy, as reported in [3, 4], respectively. Geothermal energy can also be converted to mechanical and electrical power for use in membrane desalination processes. In RO, for example, mechanical energy is utilised in the pressure pump, which provides the pressure necessary to get the feed water over the membrane. This pressure is required to remove the impurities from the feed water. When the temperature of the feed water is lowered, the viscosity of the water increases, which in turn increases the amount of pressure that must be applied during the RO process, increasing the amount of energy that must be used. Geothermal fluids with the right combination of high salinity and temperature can be used as feed water for RO systems that create freshwater. This is because the temperature tolerance of RO systems ranges from 20 °C to 35 °C. This can help supplement the amount of mechanical energy needed to pump cold water during the winter months [5]. A techno-economic analysis that performed a cost comparison between MED and RO techniques of desalination using geothermal energy was presented by [6]. The researchers concluded that the RO approach was significantly more cost effective than the MED method for the desalination of saltwater in Gulf nations.

3.3 Ocean energy potential for desalination

The term 'ocean energy' refers to the extraction of energy from the ocean in the form of heat, motion, or chemical reactions. Due to its higher energy density and reduced intermittency, it is considered a more reliable kind of renewable energy source compared to solar and wind power [8]. The mechanical energy of a wave, tide, or current is the most prevalent kind of ocean energy, and is often transformed into electrical energy through energy converters [9].

3.3.1 Ocean thermal energy

The generation of energy from heat that has already been gathered from the sun by the ocean and is sitting on its surface is known as ocean thermal energy. The temperature difference between the ocean's deeper depths and surface, which has been heated by the sun, is the source of the thermal energy released by the ocean. A French physicist named D'Arsonval is credited with making the first proposal to harness the thermal energy of the ocean in 1881, as stated by [10]. The ocean thermal energy conversion (OTEC) cycle is one way to harness energy from water. According to [11], the implementation of a thermal form of ocean energy is more straightforward by using the thermal process of desalination. Despite this, the technology is still in the research and development phase, with the limitations of a high upfront capital cost and the requirement of a large seawater pump and piping system. The authors in [12] published a study detailing the experimental design of a desalination process that utilises the heat energy of the ocean in order to create freshwater from brackish water. The findings indicate that there has been a sufficient increase in freshwater production. The result shows an adequate increase in freshwater production. The authors in [13] examined dual-use open-cycle OTEC to establish the effectiveness of the system in terms of electricity production and saltwater desalination. The low water supply rate was the main factor in determining both the pumping costs and output rates. An exergy study of syphon ash evaporation desalination was carried out by [14] making use of the thermal energy provided by the ocean. On the basis of the experiment's findings, it was determined that a sufficient amount of efficiency could be obtained with lower surface seawater, greater depth seawater, and more ash.

3.3.2 Ocean mechanical energy (wave, tidal, and current energy)

Wave energy, tidal energy, and current energy are the three components that make up the mechanical energy. In a manner analogous to the creation of energy from the wind, hydrokinetic devices, such as turbines, are utilised to capture the ocean's mechanical energy of the ocean. The pressure generated by this type of ocean energy may be utilised in pressure-driven desalination procedures in an unfiltered and undiluted form. As a result, the amount of money and energy needed for such a desalination process will be reduced. Some researchers are in favour of applying ocean mechanical energy to the RO desalination technique. This is because it is simple to employ both mechanical and electrical energy to power the RO process. The RO system was upgraded in [15] to incorporate ocean mechanical energy for the purpose of cogenerating power and desalination.

3.3.2.1 Wave energy

Wind energy is transferred to the ocean's surface as it blows from one side to another, causing the water to move in a rippling motion. This is how wave energy is created. The authors in [16] use a water hammer and a hydro-ram, respectively, to conduct their research on the technological and economic evaluation of wave-powered RO desalination. These systems not only demonstrate that it is possible to provide a direct pressure sufficient to drive the RO desalination process, but they also provide less expensive alternatives to the traditional RO system. In addition, [17] demonstrated an autonomous wave-powered RO desalination system for energy recovery.

3.3.2.2 Tidal energy

The rapid ascent and descent of water bodies create a pressure differential, providing mechanical energy which may be transformed into electrical energy. This process results in the generation of tidal energy. The gravitational pull of attraction that exists between the earth, moon, and sun, as well as the rotation of the earth and that of the moon around the planet, are all factors that contribute to the rapid rise and fall of water bodies. The authors proposed an RO desalination system using tidal energy [18], and the results indicated a 31%–41.7% decrease in the cost of producing water and an increase of 40% in the rate of water recovery compared to conventional RO systems. The cost decrease might be affected by the pressure of the tidal-RO feed. It is anticipated that the cost reduction would appreciate throughout the plant's operational lifetime.

3.3.2.3 Current energy

The motion of bodies of water in a certain direction, often controlled by the ocean's geography and the differences in temperature and salinity of the water, produces the energy known as current. This type of ocean energy has not yet been investigated in-depth, nor has it been used in the process of desalinating water to any significant degree. The feasibility of using ocean energy for RO desalination in South Africa was investigated and estimated by [19]. The researchers concluded that the RO plant powered by the energy from the ocean currents could not always provide sufficient quantities of freshwater as a unit to meet the demand of the people. This is sometimes a consequence of the insufficient water velocity.

3.3.3 Ocean chemical energy (salinity gradient energy)

The necessity of restructuring desalination plants, particularly the feed intake portion of the plants, is one of the most significant obstacles that must be overcome to implement either thermal or mechanical ocean energy into the desalination process [20]. This is one of the biggest hurdles that must be overcome. The cost of capital will increase due to this alteration. The utilisation of chemical energy, sometimes known as 'blue energy,' in the desalination of saltwater does not require an adjustment to the standard desalination process. This type of ocean energy is extracted from the combination of two sources of water, often seawater (high salty water) and river water (low saline or freshwater) with different salinity gradients under regulated settings [21]. Seawater has a high concentration of salt, whereas river water contains a low concentration of salt or freshwater. According to [21, 22], the amount of energy that can be harvested from salinity gradients worldwide is estimated to be between 2.4 and 2.6 TW, almost equivalent to the amount of power used worldwide. The difference in osmotic pressure between the two water supply sources drives the membrane desalination process. This difference is caused by the salinity gradient between the two sources of water supply. The removal of the salinity gradient energy in the form of mechanical energy for desalination may be acquired in pressure retarded osmosis, while the removal of salinity gradient energy in the form of electrical energy for desalination can be achieved in reverse electro-dialysis [23, 24]. There is still much more research to be done into how the RO desalination system can make use of the chemical energy available from the ocean. There are reports on the current state of development in [10, 25, 26].

3.4 Wind energy potential for desalination

The use of wind turbines allows for the collection of the kinetic energy created by the flow of air, the subsequent conversion of this kinetic energy into mechanical energy, and finally the generation of electrical energy. Wind power is attractive due to its low cost and excellent environmental credentials. The intermittent nature of wind energy is the main drawback to its use since it makes it difficult to maintain constant output of electricity. The integration of this technology with storage systems, the grid, or hybridisation with another kind of renewable energy source has been proposed and implemented [27] as potential solution to this problem. After the sun, wind power is one of the RES used the most frequently to power desalination operations. The use of wind energy to power desalination encompasses a variety of desalination processes and, in some cases, includes the hybridisation of desalination technologies or with another kind of renewable energy source. The RO process continues to be the most widely used technology for desalination electricity generated by wind energy. The technical and economic feasibility of wind-powered RO and MVC desalination was confirmed in [28], while [29] evaluated the effects of wind intermittency and fluctuation on the RO system and proposed three possible solutions: integration of energy storage, a hybrid energy system, or matching of RO capacity with transient energy supply. The authors in [29] propose a techno-economic evaluation of RO systems for a strategic aquifer storage and subsequent recovery (ASR) scheme as a way to lower the cost of water desalination for nations that are members of the Gulf Council Cooperation. As a case study, we looked at an ASR project that was carried out in the United Arab Emirates. The practicality of the system has been established; however, the efficient use of variable wind power will need to be mitigated by the RO plant to prevent shutting down entirely. This will result in increased plant capital costs. A wind-powered RO system for the Canary Islands was reported in [30], measuring the exergy efficiency of the system and quantifying the unit cost of freshwater. This was done to determine the cost per unit of freshwater. According to the findings, the cost per unit of freshwater is 0.76 euros per cubic metre, and the exergy efficiency of RO is more than that of wind power. The authors in [31] reviewed the economic analysis of the RO system driven by wind compared to conventional energy sources. The levelized cost of freshwater was used as the primary criterion in this comparison. The levelized cost of producing freshwater using a wind-driven RO system is lower.

3.5 Solar energy potential for desalination

The use of solar collectors or the semiconductors present in photovoltaic (PV) cells to extract energy from the Sun is the process known as solar energy. Both the sun's light and its heat may be used to create energy; specifically, thermal energy for heating and electrical energy for powering machines. Therefore, solar energy can be categorised as either PV or solar thermal [32]. Each of the two types of solar energy may be used on its own or in conjunction with another renewable energy source to power a desalination process. The intermittent nature of solar energy, much like the intermittent nature of wind energy, is the primary restriction of solar energy, which may similarly be mitigated or dealt with via hybridisation and storage systems.

3.5.1 Solar thermal-powered desalination

The solar collector is the solar thermal system component responsible for either collecting or absorbing energy from the Sun. The solar energy that the solar collector collects is first transformed into heat energy inside the absorber, and this heat energy is subsequently transmitted to the fluid that moves through the absorber [27]. This heat energy may be directly employed to power thermal desalination processes or indirectly applied to power membrane processes by converting heat energy into mechanical or electrical energy. In either way, it can be used to power desalination processes. An analysis of the key factors determining the market potential of solar thermal power desalination reveals that a dish concentrator that is combined with microgas turbines is a workable and attractive alternative to PV reverse osmosis in situations where there is a limited demand for water [33]. Despite various design improvements to improve productivity, existing designs have not yet been acceptable for large-scale production [34–36]. This was shown by developments in the utilisation of solar energy and advancements in its design for use with water desalination.

The use of concentrated solar power (CSP) to power RO is an example of an indirect application of the solar thermal system to desalination. This is a viable solution that has recently been used. CSP technology involves the steam organic Rankine cycle, dish Stirling, and other turbine and engine forms to transform the high temperature heat extracted by a regulated concentration of solar radiation in a glass mirror into either mechanical or electrical energy [37]. Dish engines, Fresnel mirror reflectors, power towers, and parabolic troughs are the four subcategories that make up CSP technology [2]. The parabolic trough is the most productive of the four subcategories. MED and RO systems are the most effective desalination processes that can be run in CSP [2]. The CSP-RO system is seen here in figure 3.2. The expense of the energy conversion unit and its installation and ongoing maintenance is the most significant disadvantage of using a solar thermal system for desalination [38].

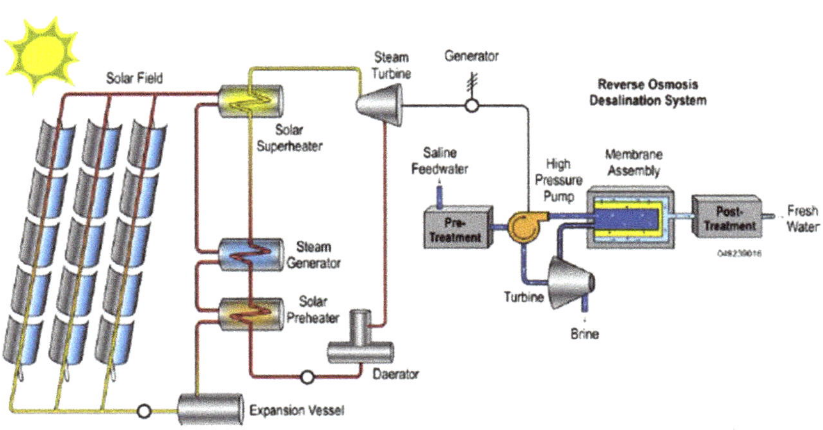

Figure 3.2. Schematic diagram of CSP reverse osmosis desalination [2, 7]. Reprinted from [7], copyright (2021) with permission from Elsevier.

3.5.2 PV-powered desalination

A solar PV energy system is based on converting solar radiation or light into electricity using equipment with a 'PV effect.' The PV effect is the indication of a voltage differential that occurs as a consequence of the light that shines on electrodes that have been systematically organised with a solid or liquid in between them [39]. Therefore, the use of solar panels with many layers is required to capture photons, which supplies sufficient energy to trigger the release of electrons from atoms in the semiconductor [32]. A solar PV system requires the use of solar panels. Electrons travel from positive to negative electrodes during this process, which results in the generation of an electric current. As a result of the unpredictable nature of solar irradiation, PV systems often use batteries for their energy storage needs. The PV-RO desalination system is shown in figure 3.3 using a schematic depiction of the system.

Since RO needs electrical energy to drive it and has become the most prominent technique of desalination, research and application of PV energy to desalination have focused more aggressively on the RO method. This is because RO requires electrical energy to drive it. In particular, as a result of this, PV-RO has become the indirect solar-powered desalination system of RES that has received the most research and has been implemented the most frequently, as depicted in figure 3.4 [40]. The authors in [41] divided articles on PV-RO based on the main idea of evaluated papers into the following four categories: influence of parameters on PV-RO system performance, economic assessment of the PV-RO unit, modification of the PV-RO system, and hybridisation of PV-RO with either another renewable energy source or a desalination technology. An in-depth analysis of the research

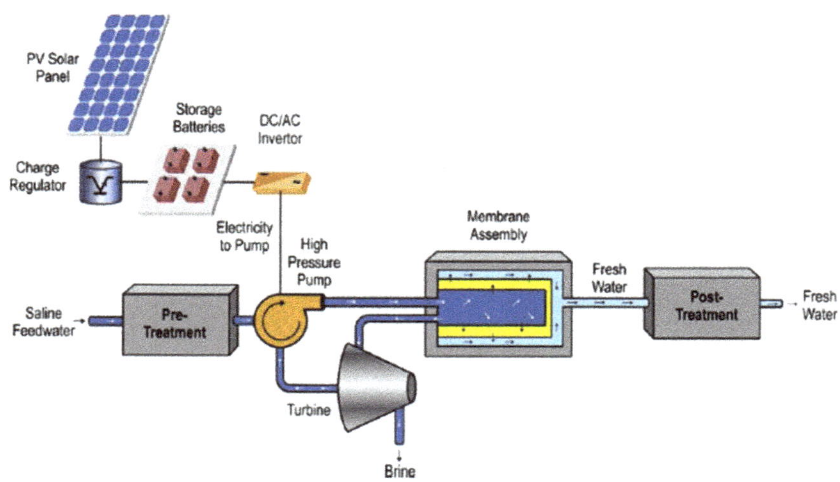

Figure 3.3. Schematic diagram of the PV reverse osmosis desalination system [2, 7]. Reprinted from [7], copyright (2021) with permission from Elsevier.

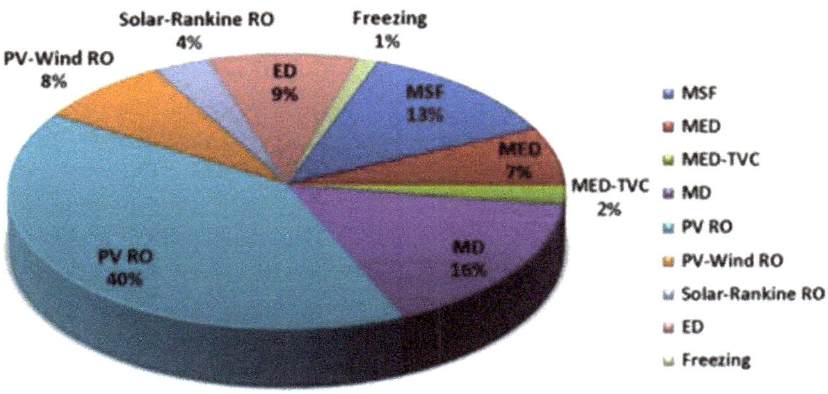

Figure 3.4. Worldwide share distribution of the installed indirect solar-powered desalination plant (2011) [7, 40]. Abbreviation: MD, membrane distillation. Reprinted from [7], copyright (2021) with permission from Elsevier.

done on PV-RO has been compiled. The primary goals of most publications that examine the influence of parameters on the performance of PV-RO systems are to research parameter adjustment regions with the aim of increasing economic efficiency and flexibility with minimal restriction of site [42, 43]. The purpose of the PV-RO modification is to increase the performance of the system so that adequate freshwater can be produced at a lower cost [44]. Eliminating the limitations posed by PVs and RO may be achieved by the hybridisation of PVs and RO with other RES or desalination techniques [45, 46]. The authors in [47, 48] provide the results of an economic analysis of PV-RO with the intention of lowering the cost of producing freshwater.

3.6 Hybrid RES for desalination

Several studies have advocated for the use of hybrid RES as a means to circumvent the drawbacks of individual RES. Because one renewable energy source may compensate for the shortcomings of the other, the hybrid system has a higher degree of dependability, increasing the system's economic efficiency and environmental effect [49]. When a conventional energy source is hybridised with RES, such as wind and solar, the conventional energy source compensates for the intermittent nature of the RES as well as their prohibitively high cost. At the same time, the use of RES reduce the harmful effects that traditional energy sources have on the surrounding environment [50, 51]. On the other hand, depending on the location and availability of RES, various RES may be hybridised to improve the cost-effectiveness and effective performance of the system [52–54]. RES may also be integrated with energy storage technologies such as capacitors, batteries, and hydrogen storage to compensate for sporadic and varying irradiation. [55–57].

3.7 Summary

The increasing demand for freshwater, coupled with concerns about traditional energy-intensive desalination methods, has led to a growing interest in utilising RES to power desalination systems. Renewable energy offers a sustainable and environmentally friendly approach to address water scarcity challenges, reducing greenhouse gas emissions, and dependence on fossil fuels. Several RES show promise in the power of desalination systems, including ocean, solar, wind, and geothermal energy.

Although ocean energy has significant potential for desalination, there are some challenges to address.

(a) Technical and engineering complexity: developing and implementing ocean energy technologies can be technically challenging and require advanced engineering solutions.
(b) Environmental impact: environmental impact assessments are crucial to ensure ocean energy projects do not harm marine ecosystems or disrupt local coastal environments.
(c) Cost and deployment: ocean energy projects often have higher upfront costs than conventional energy sources. However, costs are expected to decrease as technology advances and deployment scales up.

These challenges have kept the advancement and utilisation of ocean energy at the experimental stage so far. Despite these challenges, ocean energy for desalination presents a promising avenue for sustainable water production, contributing to global efforts to mitigate water scarcity and reduce dependence on fossil fuels.

Solar energy is one of the most abundant and readily available RES globally. It can be harnessed through PV panels to generate electricity or solar thermal collectors to produce heat. In solar-powered desalination systems, PV panels are often used to directly power RO desalination plants. Alternatively, solar thermal collectors heat a fluid to generate steam, which drives MED or MVC desalination processes. Solar energy offers a decentralised and sustainable solution, particularly in arid regions with high solar irradiance and water scarcity.

Wind energy is another abundant and predictable renewable resource that can be used to power desalination systems. Wind turbines convert the kinetic energy of the wind into electricity which can then be used to operate various desalination technologies. Wind-powered desalination systems often integrate with RO or electrodialysis processes, where excess energy can be stored in batteries or used in other applications. Wind energy variability can be managed through energy storage systems or by integrating multiple renewable sources to ensure continuous desalination operations.

Geothermal energy derives from the heat stored in the Earth's crust and is a reliable and sustainable source of power. Geothermal-powered desalination systems utilise geothermal heat to drive desalination processes such as MED or MVC. Geothermal energy is especially suitable for coastal regions, where the proximity of geothermal resources to seawater sources simplifies the integration of the desalination plant.

In many cases, a single renewable energy source may not provide a consistent and reliable power supply to meet the demands of a desalination plant. To overcome this challenge, hybrid renewable energy systems are gaining popularity. These systems combine two or more RES, such as solar and wind, or solar and geothermal, to ensure a continuous and stable energy supply for desalination operations. Hybrid systems offer enhanced reliability, energy efficiency, and cost-effectiveness.

RES present promising opportunities for the powering of desalination systems in a sustainable and environmentally friendly manner. By utilising renewable resources, these desalination plants can reduce their carbon footprint, minimise dependency on finite fossil fuels, and contribute to global efforts to combat water scarcity and climate change. As technology advances and the cost of renewable energy continues to decline, the integration of RES in desalination systems is expected to play a pivotal role in building a more resilient and sustainable future of water.

References

[1] Abdelkareem M A, El Haj Assad M, Sayed E T and Soudan B 2018 Recent progress in the use of renewable energy sources to power water desalination plants *Desalination* **435** 97–113

[2] Al-Karaghouli A and Kazmerski L L 2013 Energy consumption and water production cost of conventional and renewable energy-powered desalination processes *Renew. Sustain. Energy Rev.* **24** 343–56

[3] Bourouni K, Deronzier J C and Tadrist L 1999 Experimentation and modelling of an innovative geothermal desalination unit *Desalination* **125** 147–53

[4] Bouguecha S and Dhahbi M 2003 Fluidised bed crystalliser and air gap membrane distillation as a solution to geothermal water desalination *Desalination* **152** 237–44

[5] Gude V G 2016 Geothermal source potential for water desalination - current status and future perspective *Renew. Sustain. Energy Rev.* **57** 1038–65

[6] Loutatidou S and Arafat H A 2015 Techno-economic analysis of MED and RO desalination powered by low-enthalpy geothermal energy *Desalination* **365** 277–92

[7] Okampo E J and Nwulu N 2021 Optimisation of renewable energy powered reverse osmosis desalination systems: a state-of-the-art review *Renew. Sustain. Energy Rev.* **140** 110712

[8] Lehmann M, Karimpour F, Goudey C A, Jacobson P T and Alam M R 2017 Ocean wave energy in the United States: current status and future perspectives *Renew. Sustain. Energy Rev.* **74** 1300–13

[9] Bilgili M, Ozbek A, Sahin B and Kahraman A 2015 An overview of renewable electric power capacity and progress in new technologies in the world *Renew. Sustain. Energy Rev.* **49** 323–34

[10] Li Z, Siddiqi A, Anadon L D and Narayanamurti V 2018 Towards sustainability in water-energy nexus: ocean energy for seawater desalination *Renew. Sustain. Energy Rev.* **82** 3833–47

[11] Sawyer R A and Maratos D F 2001 An investigation into the economic feasibility of unsteady incompressible duct flow (waterhammer) to create hydrostatic pressure for seawater desalination using reverse osmosis *Desalination* **138** 307–17

[12] Senthil Kumar R, Mani A and Kumaraswamy S 2007 Experimental studies on desalination system for ocean thermal energy utilisation *Desalination* **207** 1–8

[13] Kim A S, Kim H J, Lee H S and Cha S 2016 Dual-use open cycle ocean thermal energy conversion (OC-OTEC) using multiple condensers for adjustable power generation and seawater desalination *Renew. Energy* **85** 344–58
[14] jiang Jin Z, Ye H, Wang H, Li H and yuan Qian J 2017 Thermodynamic analysis of siphon flash evaporation desalination system using ocean thermal energy *Energy Convers. Manag.* **136** 66–77
[15] Ferreira R and Estefen S 2011 Ocean power conversion for electricity generation and desalinated water production *World Renewable Energy Congress* 57 *(Linköping, Sweden)* 2198–205
[16] Maratos D F 2003 Technical feasibility of wave-power for seawater desalination using the hydro-ram (Hydram) *Desalination* **153** 287–93
[17] Folley M, Peñate Suarez B and Whittaker T 2008 An autonomous wave-powered desalination system *Desalination* **220** 412–21
[18] Ling C, Wang Y, Min C and Zhang Y 2018 Economic evaluation of reverse osmosis desalination system coupled with tidal energy *Front. Energy* **12** 297–304
[19] Leijon J, Forslund J, Thomas K and Boström C 2018 Marine current energy converters to power a reverse osmosis desalination plant *Energies* **11** 2880
[20] Alvarez-Silva O A 2017 *Implementing Salinity Gradient Energy at River Mouths* (New York: Elsevier)
[21] Jia Z, Wang B, Song S and Fan Y 2014 Blue energy: current technologies for sustainable power generation from water salinity gradient *Renew. Sustain. Energy Rev.* **31** 91–100
[22] Sales B B, Saakes M, Post J W, Buisman C J N, Biesheuvel P M and Hamelers H V M 2010 Direct power production from a water salinity difference in a membrane-modified supercapacitor flow cell *Environ. Sci. Technol.* **44** 5661–5
[23] La Cerva M L *et al* 2017 Coupling CFD with a one-dimensional model to predict the performance of reverse electrodialysis stacks *J. Memb. Sci.* **541** 595–610
[24] Straub A P, Deshmukh A and Elimelech M 2016 Pressure-retarded osmosis for power generation from salinity gradients: is it viable? *Energy Environ. Sci.* **9** 31–48
[25] Gu H, Sun X, Li W, Lin Y, Ma S and Liu H 2010 A review on the development of tidal current energy in China *Renew. Sustain. Energy Rev.* **15** 1141–6
[26] Bane J *et al* 2017 The economics of electricity generation from Gulf Stream currents *Energy* **134** 649–58
[27] Lai W, Ma Q, Lu H, Weng S, Fan J and Fang H 2016 Effects of wind intermittence and fluctuation on reverse osmosis desalination process and solution strategies *Desalination* **395** 17–27
[28] Baldea M *et al* 2007 Feasibility study on wind-powered desalination *Desalination* **203** 463–70
[29] Romero-Ternero V, García-Rodríguez L and Gómez-Camacho C 2005 Thermoeconomic analysis of wind-powered seawater reverse osmosis desalination in the Canary Islands *Desalination* **186** 291–8
[30] García-Rodríguez L, Romero-Ternero V and Gómez-Camacho C 2001 Economic analysis of wind-powered desalination *Desalination* **137** 259–65
[31] Rosales-Asensio E, Borge-Diez D, Pérez-Hoyos A and Colmenar-Santos A 2019 Reduction of water cost for an existing wind-energy-based desalination scheme: a preliminary configuration *Energy* **167** 548–60
[32] Melián G, Peñate B, de la Fuente J A, García-Rodríguez L and Delgado-Torres A M 2019 Water desalination by solar-powered RO systems *Current Trends and Future Developments in*

(Bio-) Membranes: Renewable Energy Integrated with Membrane Operations ed A Basile, A Cassano and A Figoli (New York: Elsevier) pp 45–84
[33] Pouyfaucon A B and García-Rodríguez L 2018 Solar thermal-powered desalination: a viable solution for a potential market *Desalination* **435** 60–9
[34] Li C, Goswami Y and Stefanakos E 2013 Solar assisted seawater desalination: a review *Renew. Sustain. Energy Rev.* **19** 136–63
[35] El-Sebaii A A and El-Bialy E 2015 Advanced designs of solar desalination systems: a review *Renew. Sustain. Energy Rev.* **49** 1198–212
[36] Reif J H and Alhalabi W 2015 Solar-thermal powered desalination: its significant challenges and potential *Renew. Sustain. Energy Rev.* **48** 152–65
[37] Palenzuela P, Zaragoza G, Alarcón-Padilla D C, Guillén E, Ibarra M and Blanco J 2011 Assessment of different configurations for combined parabolic-trough (PT) solar power and desalination plants in arid regions *Energy* **36** 4950–8
[38] Ahmed F E, Hashaikeh R and Hilal N 2019 Solar powered desalination—Technology, energy and future outlook *Desalination* **453** 54–76
[39] Goetzberger A, Hebling C and Schock H W 2003 Photovoltaic materials, history, status and outlook *Mater. Sci. Eng. R Rep.* **40** 1–46
[40] Ali M T, Fath H E S and Armstrong P R 2011 A comprehensive techno-economical review of indirect solar desalination *Renew. Sustain. Energy Rev.* **15** 4187–99
[41] Kasaeian A, Rajaee F and Yan W M 2019 Osmotic desalination by solar energy: a critical review *Renew. Energy* **134** 1473–90
[42] Freire-Gormaly M and Bilton A M 2018 Experimental quantification of the effect of intermittent operation on membrane performance of solar powered reverse osmosis desalination systems *Desalination* **435** 188–97
[43] Lee S, Myung S, Hong J and Har D 2016 Reverse osmosis desalination process optimized for maximum permeate production with renewable energy *Desalination* **398** 133–43
[44] Kelley L C and Dubowsky S 2013 Thermal control to maximize photovoltaic powered reverse osmosis desalination systems productivity *Desalination* **314** 10–9
[45] Garg M C and Joshi H 2014 Optimization and economic analysis of small-scale nano-filtration and reverse osmosis brackish water system powered by photovoltaics *Desalination* **353** 57–74
[46] He W, Wang Y and Shaheed M H 2015 Stand-alone seawater RO (reverse osmosis) desalination powered by PV (photovoltaic) and PRO (pressure retarded osmosis) *Energy* **86** 423–35
[47] Lamei A, van der Zaag P and von Münch E 2008 Impact of solar energy cost on water production cost of seawater desalination plants in Egypt *Energy Policy* **36** 1748–56
[48] Poullikkas A 2010 An optimization model for the production of desalinated water using photovoltaic systems *Desalination* **258** 100–5
[49] Bajpai P and Dash V 2012 Hybrid renewable energy systems for power generation in stand-alone applications: a review *Renew. Sustain. Energy Rev.* **16** 2926–39
[50] Abdelshafy A M, Hassan H and Jurasz J 2018 Optimal design of a grid-connected desalination plant powered by renewable energy resources using a hybrid PSO–GWO approach *Energy Convers. Manag.* **173** 331–47
[51] Xuan T D, Alikulov K, Nakagoshi N, Aminov Z and Higashi O 2016 Analysis of environmental effect of hybrid solar-assisted desalination cycle in Sirdarya Thermal Power Plant, Uzbekistan *Appl. Therm. Eng.* **111** 894–902

[52] Narayan G P, McGovern R K, Zubair S M and Lienhard J H 2012 High-temperature-steam-driven, varied-pressure, humidification-dehumidification system coupled with reverse osmosis for energy-efficient seawater desalination *Energy* **37** 482–93

[53] Filippini G, Al-Obaidi M A, Manenti F and Mujtaba I M 2018 Performance analysis of hybrid system of multi effect distillation and reverse osmosis for seawater desalination via modelling and simulation *Desalination* **448** 21–35

[54] Minhas M B, Jande Y A C and Kim W S 2014 Combined reverse osmosis and constant-current operated capacitive deionization system for seawater desalination *Desalination* **344** 299–305

[55] Park G L, Schäfer A I and Richards B S 2013 Renewable energy-powered membrane technology: supercapacitors for buffering resource fluctuations in a wind-powered membrane system for brackish water desalination *Renew. Energy* **50** 126–35

[56] Korpås M and Greiner C J 2008 Opportunities for hydrogen production in connection with wind power in weak grids *Renew. Energy* **33** 1199–218

[57] Ayodele T R and Ogunjuyigbe A S O 2015 Mitigation of wind power intermittency: storage technology approach *Renew. Sustain. Energy Rev.* **44** 447–56

IOP Publishing

Optimisation of Renewable Energy Powered Desalination Systems
A sustainable techno-economic and environmental consideration
Ewaoche John Okampo, Nnamdi Nwulu and Pitshou Ntambu Bokoro

Chapter 4

Modelling and simulation in energy and water systems: advancements and challenges

This chapter reviews the evolving landscape of these critical domains, exploring the advancements, methodologies, and challenges associated with modelling and simulation techniques. This comprehensive study navigates through the multifaceted applications of modelling and simulation in energy and water systems. It examines how these techniques aid in predicting, analysing, and optimising resource interactions, operational processes, and policy interventions. It encompasses advancements in computational tools, data analytics, and machine learning algorithms, which have revolutionised the accuracy and scope of modelling. It explores the integration of real-time data, remote sensing, and geographic information systems to enhance the fidelity of simulations, enabling more precise predictions of resource availability and demand. From renewable energy integration to water desalination, this chapter examines the pivotal role of modelling in optimising energy–water systems. Furthermore, the chapter identifies research trends and advancement in the area of optimal renewable energy integration with reverse osmosis desalination. It is recommended that in the future, modelling and simulation of this integrated system includes a combined optimisation strategy, which takes into account the size, operation, and thermodynamic impact of the system to achieve a more comprehensive level of results.

4.1 Introduction

The intertwined challenges of meeting the increasing energy demand and ensuring sustainable water management have emerged as critical concerns in the face of global population growth, urbanisation, and environmental changes. As the world seeks to strike a delicate balance between economic growth and environmental preservation, the role of advanced modelling and simulation techniques becomes

doi:10.1088/978-0-7503-6154-5ch4

increasingly indispensable. In recent years, significant progress has been made in developing sophisticated computational tools that enable researchers and policy-makers to better understand the complex interplay between energy generation, distribution, and water resource management.

Modelling involves creating a simplified representation of a real-world system, process, or phenomenon. These representations can take different forms, such as mathematical equations, physical replicas, computer programmes, or conceptual diagrams. The purpose of modelling is to capture the essential features and dynamics of the real system while abstracting away unnecessary details [1]. Models can be of different types: (i) physical models: physical objects or prototypes that resemble the real system, used to study specific aspects of the system's behaviour; (ii) mathematical models: equations and mathematical expressions that describe the relationships and interactions between the components of the system; and (iii) computer models: digital simulations created using software, often based on mathematical models, to mimic the behaviour of the real system [2]. The level of detail and complexity in a model depends on the specific objectives and constraints of the study. A good model should strike a balance between simplicity and accuracy, allowing researchers to gain insight and make predictions about the real-world system. Simulation, on the other hand, is the process of running a model over time to observe and analyse its behaviour. By manipulating the model's parameters and initial conditions, researchers can investigate how the system responds under different scenarios. Simulation allows us to study the system's dynamics, understand its performance, and identify potential issues or opportunities. It can be classified into two main types [3]. Continuous simulation: in this type of simulation, the system's behaviour is modelled as continuous changes over time. Examples include the motion of planets, weather patterns, or fluid flow in pipes. Discrete event simulation: this type of simulation focusses on modelling the system as a series of discrete events that occur at specific points in time. Examples include computer networks, manufacturing processes, and queuing systems.

In general, modelling and simulation play a vital role in advancing knowledge, making informed decisions, and mitigating risks in numerous domains. They allow researchers to explore different scenarios without affecting the real system, saving time, saving resources, and often providing valuable insights that might be challenging to obtain through direct experimentation alone.

The concept of employing models to predict and analyse energy and water systems is not new, but the convergence of advanced computational methods, big data analytics, and improved understanding of the underlying physical processes has revolutionised the field [4]. This interdisciplinary endeavour uses principles from engineering, environmental science, economics, and other domains to construct comprehensive models capable of simulating and optimising various aspects of energy and water systems [5]. The results of such research carry profound implications for the design of sustainable policies and the advancement of innovative technologies that can pave the way for a more resilient and ecologically balanced future [6].

As we delve into the intricacies of modelling and simulation in energy and water, this review article aims to present a comprehensive analysis of the latest advancements, methodologies, and challenges in the field. The synthesis of existing research will shed light on the various approaches adopted by researchers, the integration helping tackle issues related to renewable energy, water distribution networks, climate change, and the optimisation of resource allocation. This chapter is structured as follows. In the following section, we will delve into the foundational principles of modelling and simulation in the context of energy and water systems. We will then examine recent case studies that demonstrate the application of these techniques in real-world scenarios. Furthermore, the article will critically assess the strengths and limitations of existing models, exploring opportunities for future developments and enhancements to bridge the gap between theory and practise.

4.1.1 Energy modelling and simulation

The following are types of energy system models:

- Optimisation models
- Techno-economic models
- Integrated assessment models

These models are applied in areas such as forecasting energy demand; energy supply chain—energy generation, transmission [7], and distribution planning; renewable energy and grid modelling; energy policies; and scenario planning. Simulation of energy supply and generation systems, including fossil fuels and renewables, allows the evaluation of different energy generation scenarios. These models help identify the most efficient and cost-effective mix of energy sources to meet demand while minimising environmental impact. Their analysis helps optimise grid operation, enhance reliability, and identify potential vulnerabilities or bottlenecks [5]. Furthermore, modelling and simulation play a crucial role in the integration of intermittent renewable energy sources into the grid. Such models enable the evaluation of energy storage technologies [8], demand response strategies, and grid management techniques [9] to improve the reliability and stability of the integration of renewable energy [10]. By using energy models, policymakers can assess the implications of various energy policies, such as carbon pricing, emission reduction goals, and incentives for renewable energy adoption [11]. These models help to understand the environmental impact of different energy scenarios and identify pathways towards sustainable development. Taking advantage of the power of data analysis, machine learning, and optimisation techniques, energy modelling and simulation continue to evolve and play an instrumental role in shaping the future of the energy sector [4]. These tools contribute to informed decision making, improved resource management, and the transition to a more sustainable and resilient energy system.

4.1.2 Water modelling and simulation

The following are types of water system models:

- Hydrological models.
- Hydraulic models.
- Water quality models.
- Water network models.

- Water treatment and reuse models.
- Water desalination models.

These models are applied in areas such as water resource management and planning, flood forecasting and management, drought analysis and mitigation, water infrastructure design and optimisation, and water treatment and purification. Some research in water modelling and simulation discusses the use of thermodynamic modelling to analyse energy efficiency, entropy production, and exergy losses in different desalination processes, such as reverse osmosis, multi-stage flash distillation, and membrane distillation [12, 13]. Others use the principle of heat mass transfer to explore how these models help to understand the thermal and concentration distributions within desalination units, thereby guiding design improvement and improving energy recovery strategies [14]. With the proliferation of membrane-based desalination technologies, the authors in [15] examine the modelling of membrane transport and separation processes. It discusses techniques such as transport models, pore-scale modelling, and molecular dynamics simulations to predict membrane performance, fouling behaviour, and long-term durability [16, 17]. Furthermore, since desalination systems involve complex interactions of various physical and chemical processes, multi-physics and multi-scale modelling play a crucial role in gaining comprehensive insights into the behaviour of the system [18]. Water modelling explores the integration of multiple modelling approaches to capture the interplay between thermal, hydrodynamic, and chemical phenomena.

4.1.3 Integrated energy–water modelling and simulation

Integrated energy–water modelling and simulation is a specialised approach that considers the interdependencies between energy and water systems within a given region or area [19]. It involves creating comprehensive models that capture the complex interactions between energy production, consumption, and water use [20]. The energy section of this integrated system focusses on understanding the production, distribution, and consumption of various energy sources within the system. It includes modelling power plants, renewable energy installations (e.g. solar farms, wind turbines), energy storage facilities, and the grid infrastructure. The energy model estimates the demand for electricity, heating, cooling, and other energy services based on historical data and future projections. The water component of the model accounts for the sources, uses, and distribution of water resources. It includes modelling natural water systems such as rivers, lakes, and seas, as well as human activities such as water supply, irrigation, industrial processes, wastewater treatment, and desalination. Water availability and demand are critical factors in this part of the model. Integrated models are expected to capture the feedback loops and interdependencies between energy and water systems [21]. For example, power generation often requires significant water resources for cooling (e.g. in thermal power plants), while water treatment and distribution rely on energy input. Changes in water availability due to climate patterns or water use can affect energy production and vice versa. Understanding these connections is crucial for a holistic analysis.

This integrated approach allows researchers, policymakers, and stakeholders to assess the sustainability, efficiency, and potential trade-offs between energy and water resources [22]. It also allows the evaluation of the impact of different scenarios and policies in energy and water sectors. For instance, they can assess the effects of increased renewable energy deployment on water consumption or the implications of water conservation measures on energy demand. This analysis helps to make informed decisions to optimise resource management and ensure long-term sustainability. In addition, integrated models can assess the vulnerabilities of energy and water systems to the impacts of climate change, such as changing precipitation patterns, sea level rise, and extreme weather events. This analysis guides resilience planning and adaptation strategies.

The subsequent sections of this chapter and the following chapters of this book explore the ability of the integrated model to aid in optimising the design and operation of energy and water infrastructure projects, reducing inefficiencies, and minimising environmental impacts. Emphasis is placed on the application of optimisation techniques to integrated renewable energy–powered reverse osmosis desalination systems.

4.2 Optimisation of renewable energy source—reverse osmosis desalination system

System optimisation is the process of minimising or maximising specific system parameters to improve the general efficiency. This is accomplished by selecting the best feasible variable that satisfies the system constraints in order to minimise or maximise certain system parameters. The nature of the design variables determines whether an optimisation issue is characterised as static or dynamic (parameter or trajectory), and the existence of constraints determines whether the problem is constrained or unconstrained. If the problem is underlying the physical structure, then it might be either optimum or non-optimal control. The optimisation equation might be linear programming, non-linear programming, quadratic programming, or geometric programming. Permissible values of design variables could be integer programming or non-integer programming. Furthermore, when an optimisation is based on the capacity to separate functions, it could inform of programming that can be separated or not separated. Also, it can be classified on a number of objective functions, which could be single or multi-objective problems. In addition, the deterministic nature of variables can be a measure of type of optimisation, for example, stochastic (with no certain variables) or deterministic (with known parameters or variables) [23].

Renewable energy sources (RES)-powered desalination optimisation primarily comprises a mixture of many optimisation issues, which often include optimum control, non-linear optimisation, constrained optimisation, and deterministic optimisation. Due to the complexity of the combination of various optimisation issues, it is imperative that great consideration be given to the selection of suitable optimisation methods to use as a tool to resolve the difficulties. Mathematical approaches, also known as classical or analytical optimisation techniques, and artificial intelligence (AI) techniques, which are heuristic or metaheuristic techniques, are the two categories into which optimisation solution techniques can be categorised according to figure 4.1 [25].

The traditional procedures rely on a selection of mathematical methods, such as knowledge of the gradient of functions to address optimisation issues. These approaches are well known for their capacity to produce an accurate solution to a linear or convex optimisation problem. This ability has helped these methods gain widespread recognition. Problems that are non-linear or non-convex are notoriously difficult to solve in a way that yields the best possible results. On the other hand, AI methods have recently emerged as one of the most popular approaches to solving optimisation issues. Computer-aided heuristic solvers, which give an optimum solution by using an iterative methodology, have AI algorithms incorporated into them. Therefore, metaheuristic approaches offer the benefit of quick computing, but the answers they provide are approximative rather than accurate [26].

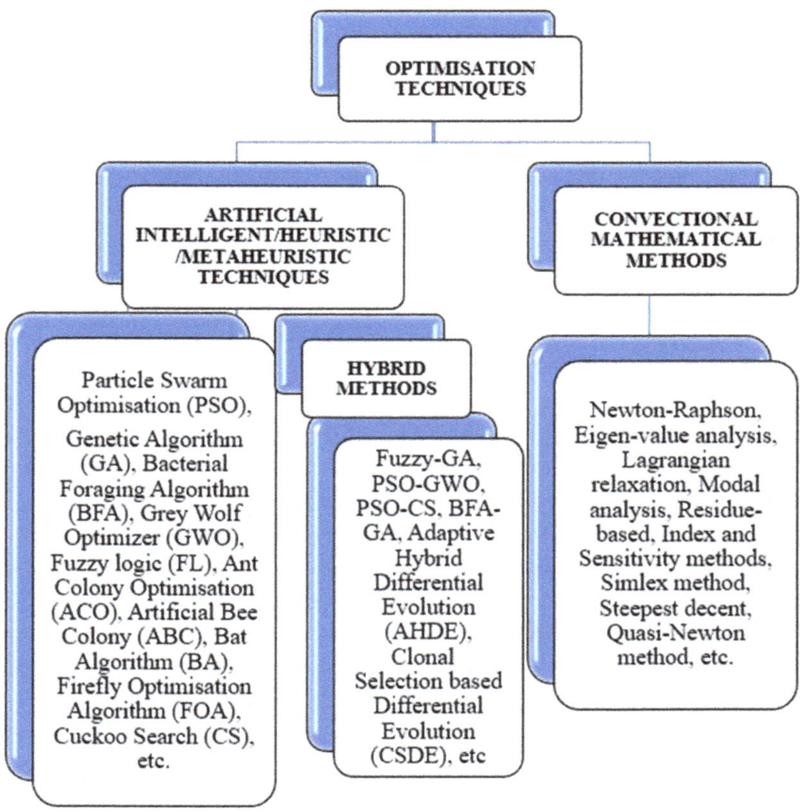

Figure 4.1. Classification of optimisation solution techniques [24]. Reprinted from [69] CC BY 4.0.

An objective function is required for each optimisation task. Some performance indicators are used to determine and assess whether a goal was accomplished. Therefore, performance metrics determine the construction of an appropriate optimisation problem. In the case of optimisation problems involving RES-powered desalination, the performance indicators that are most frequently used in the formation of objective functions are as follows: (i) system resiliency, (ii) energy requirement and consumption by the desalination process, (iii) cost of production of freshwater using various cost matrices, (iv) volume of the freshwater output, (v) brine handling and treatment, (vi) environmental impacts of both energy sources and the desalination processes, (vii) permeate flux rate, and (viii) system life span. Therefore, the finest optimum desalination system will be one that has extremely low energy needs and consumption, a high permeate flow rate, a large volume of freshwater production, a low risk of environmental effects, a long life span, and a minimum production cost. All these characteristics should be present in the ideal system. Due to certain technological and physical limitations, it may be difficult to get the best possible results from all these performance assessments in a system at the same time in a real application. However, to solve a single objective optimisation issue, improving a desalination system to achieve one of the performance indicators would be necessary. A multi-objective optimisation issue will be defined and addressed whenever there is a combination of more than one performance indicator. In any event, the system is anticipated to provide improved performance despite it physical and technological limitations. Due to the stochastic character of certain RES, reliability assessments should be done to ensure that RES–reverse osmosis (RO) desalination systems work adequately. Dependability indicators such as the chance of power loss, the probability of load loss, and the predicted amount of energy that is not delivered can be used to evaluate the system's dependability [27], while several cost matrices are used to analyse the economic feasibility of an RES-RO desalination system. The total cost of the system, the total annualised cost of the system (ACS), the levelized cost of electricity (LCOE), the net present cost, and the total cost of the life cycle are some of the cost matrices that may be utilised. The mathematical representation of the performance indicators can be found in table 4.1. This table offers a variety of cost matrices and reliability indicators.

In more recent research, certain applications of optimisation models to optimise the RO desalination process have been suggested. Evaluation of some of these studies will be based on three different optimisation methodologies: system sizing, operational optimisation, and thermodynamic optimisation. The primary notions obtained from the analysed literature served as the basis for these three different optimisation strategies. The next subsections will provide more specific information on some of the applications of the three different optimisation strategies to RES-RO systems.

4.2.1 System sizing optimisation

Sizing a system involves developing an appropriate design for a significant amount of system capacity while adhering to the restrictions imposed by the system's components. The authors in [36] modelled six grid-independent hybrid renewable

Table 4.1. Simplified mathematical model of some performance indicators [24]. Reprinted from [24], copyright (2021) with permission from Elsevier.

Performance indicators	Cost matrices	Mathematical equations	References
Economic indicators	Capital recovery factor (CRF)	$\text{CRF} = \left[\frac{i(i+1)^n}{(i+1)^n - 1}\right]$	[28, 29]
	System total cost	$\text{STC} = C_{\text{capital}} + C_{\text{operation}} + C_{\text{maintenance}} + C_{\text{installation}} + C_{\text{replacement}}$	[27, 30]
	Total annualised cost of the system	$\text{ACS} = C_{\text{capital}}(\text{CRF}) + C_{\text{replacement}}(\text{CRF}) + C_{\text{maintenance}}$	[27, 31]
	Levelized cost of electricity	$\text{LCOE} = \frac{\text{ACS + energy consumption cost}}{\text{annual energy production}}$	[32, 33]
	Net present value	$\text{NPV} = \sum_{n=1}^{N} \frac{\text{investment in year } n\ Q_n}{(1 + \text{return rate } r)^n} - 1$	[27, 34]
	Life cycle cost	$\text{LCC} = \sum_{n=1\ldots N}^{N} \text{NPV}$	[34–36]
	Cost of water production	$\text{COW}\left(\frac{\$}{\text{m}^3}\right) = \frac{\left(\frac{\text{Investment cost}}{\text{Capital cost recovery factor}}\right) + \text{Cost of operation \& management}}{\text{volume of water / year}}$	[28, 37]
	Gas emission cost	$\text{GEC}\left(\frac{\$}{\text{m}^3}\right) = \text{Electricity requirement} \left(\frac{\text{kWh}}{\text{m}^3}\right) \times \text{Emission factor}\left(\frac{\text{kgCO}_{2-e}}{\text{kWh}}\right) \times \text{Carbon tax}\left(\frac{\$}{\text{kgCO}_{2-e}}\right)$	[37]
Reliability indicators	Specific energy consumption	$\text{SEC} = \frac{\text{power consumed by pump } (W_{\text{pump}})}{\text{permeateflux rate}}$	[38, 39]
	Loss of power supply probability	$\text{LPSP} = \frac{\sum_{t=1}^{T} \text{power not supplied}(t)}{\Sigma_{t=1}^{T} E_{\text{load}}(t)}$	[27, 36]
	Loss of load probability	$\text{LLP} = \frac{\Sigma_{t=1}^{T} \begin{cases} 0, & L(t) > P(t) \\ 1, & L(t) \leqslant P(t) \end{cases}}{T}$	[27]

energy systems to power RO desalination systems. These systems include solar-battery, solar-hydrogen, wind-battery, wind-hydrogen, solar-wind-battery, and solar-wind-hydrogen. The authors' goal was to minimise total life cycle cost for each system by using an improved bee algorithm and comparing it to a harmony search algorithm. The hybridisation with battery energy storage was determined to be more cost-effective than the storage of hydrogen energy storage, while the solar battery–powered RO unit was the most cost-effective energy system of the six that were tested. A grid-connected RES-RO desalination facility in the city of Alexandria, Egypt, was optimised by reducing total investment cost and CO_2 using a hybrid of PSO and GWO [40]. This was achieved by optimising the plant to use the most efficient combination of RES and RO. The findings also demonstrate that battery storage is more cost-effective than a hydrogen storage system, and the hybrid PSO-GWO storage solution proved to be more efficient than either PSO or GWO on its own. The authors used a GA [30] to create an optimal desalination model powered by photovoltaic and wind energy for a small town and a residential family. The goal of the model was to find the ideal number and kind of system units at the smallest overall cost. An overview of the size optimisation models currently being looked at may be found in table 4.2.

4.2.2 Operational optimisation and control

This section highlights the significance of optimisation and control strategies in energy-efficient desalination. It discusses using dynamic modelling and control algorithms to maximise system performance while minimising energy consumption and operating costs. To acquire the best possible results from the RO desalination system, several components of the system might have their configurations changed or their settings adjusted. In many instances, studies of operational optimisation focused primarily on studying the system's performance rather than giving much thought to the scale of the operation. In [33], the levelized unit cost of water and energy for a wind-powered seawater RO system on the island of Gokceada in Turkey was analysed. The energy need, the study of emission reduction, and the prediction of an ideal levelized cost for the system were all carried out using an RO system analysis, or ROSA, model. According to the findings, the predicted levelized cost of water (LCOW) for the system's energy requirement with a capacity range of 6–30 kW, which is between USD 2.96 and 6.457 per cubic metre, and the predicted LCOE for the system is between USD 0.077 and 0.155 per kilowatt-hour when using a wind turbine with a capacity of 30 kW and an 8% discount rate based on the turbine's specific cost. The study anticipated a yearly decrease in CO_2 emissions of 80 028 tons when RO is installed along a wind turbine with a capacity of 30 kW. The possibility of utilising water and hydrogen as double storage for a continuous supply of energy and water from a PV-wind–powered RO system at a low cost was studied in [47]. This study looked at the feasibility of combining water and hydrogen as double storage for energy and water. HOMER and ROSA successfully optimised the system. A significant reduction in CO_2 emissions was also accomplished in addition to verifying the system's viability and dependability from a technical

Table 4.2. Summary of sizing optimisation models of RES desalination plants [24]. Reprinted from [24], copyright (2021) with permission from Elsevier.

Objective function	Constraints	Optimisation techniques	Outlook	References
Minimisation of total life cycle cost	Inequality constraints: bound limit of the area of the photovoltaic (PV), the area of wind turbine, the number of hydrogen and battery storage systems, the energy stored, and the state of charge. Equality constraint: loss of power supply.	Improved bee algorithm compared to harmony search (HS) algorithm	The hybridisation of a solar power, battery, and RO system is considered the most cost-effective system compared to five other configurations.	[36]
Minimisation of total life cycle cost	Inequality constraints: bound limit of the area of the PV, the area of wind turbine, the number of hydrogen and battery storages, the energy stored, and the state of charge. Equality constraint: Loss of power supply.	HS algorithm and HS-based chaotic search (HSBCS)	The proposed HSBCS yields a better result than HS algorithm	[34]
Minimisation of total life cycle cost	Inequality constraints: bound limit of the area of the PV, the area of wind turbine, the number of hydrogen and battery storage systems, the energy stored, and the state of charge. Equality constraint: loss of power supply.	Artificial bee swarm optimisation (ABSO)	ABSO proves to be effective. The hybrid system led to a cost-effective system for different maximum loss of power supply probability (0%–100%).	[41]

Minimisation of total life cycle cost with loss of power probability as reliability index	Inequality constraints: bound limit of the area of the PV, the area of wind turbine, the number of hydrogen and battery storage systems, the energy stored, and the state of charge. Equality constraint: loss of power supply.	PSO, ABSO, simulated annealing, HS, chaotic search, Tabu search algorithm	The hybrid metaheuristic improved HS-based chaotic simulated annealing yielded a better result than eight other metaheuristic algorithms, while ABSO yielded the worst performance.	[35].
Minimisation of the total investment cost and CO_2 emissions	Inequality constraints: bound limit of the number of PV arrays, wind turbines, hydrogen tank and battery storages, electrolyser, inverters, fuel cell, and capacity of RO plant and water tank—all are 0. Renewable factor 70%. Limit of diesel power generation.	Hybrid PSO-GWO	The result also shows that battery storage is more cost-effective than the hydrogen storage system, and the hybrid PSO-GWO proved more effective than either PSO or GWO.	[42]
Determination of the optimal configurations of the hybrid system at minimal cost	Inequality constraints: the state of charge of hydrogen storage. Equality constraints: power balance and water flow balance equations.	A new optimisation iterative algorithm was formulated using MATLAB software	The formulation method can be applied to any desalination method. The result shows improvement in energy supply reliability and reduced energy storage requirements, and consequently reduced system installation cost.	[43]

(Continued)

Table 4.2. (*Continued*)

Objective function	Constraints	Optimisation techniques	Outlook	References
Determination of the optimal number and type of unit of the system at minimal total system cost	Inequality constraints: limit of numbers of PV modules, wind power generation blocks, battery storage systems, RO plant, and PV tilt angle.	GA	The overall configuration is profoundly affected by the capital cost. The hybrid of PVs and wind generator is more cost-effective than PVs or wind stand-alone.	[30]
Minimisation of excess energy at minimal cost	Operating reserve of 10%, 15% and 25% of the load demands, output wind and solar power, respectively.	An optimisation algorithm was developed to select the best combination of hybrid RES.	The impact of the optimisation method on the unit cost of energy and, consequently, the unit cost of desalinated water was demonstrated.	[44, 45]
Minimisation of total capital cost (capital + operational cost)	Daily state of the tank and maximum permeable depth of battery discharge.	GA	Only an economic estimation of power provided by RES was presented using the hybrid optimisation model for electric renewables (HOMER). The hybrid of PVs, wind, and battery shows significant improvement in the system.	[46]

perspective; [48] describes the hybridisation of PVs and concentrated solar power–organic Rankine cycle (ORC) to power multi-effect distillation (MED)-RO to minimise the unit operating cost. The use of formulated mixed integral linear programming, or MILP, accomplishes this. Accuracy and speed may be shown using a rolling horizon technique in conjunction with a 24 h unit commitment issue defined as MILP. Both dependability and sustainability are increased thanks to the hybrid system. Table 4.3 summarises further operational optimisation that may be applied to RES-RO systems.

4.2.2.1 Thermodynamic optimisation

Heat and electrical energy transfer, as well as their connection to temperature and the mass of water flow in an RO desalination system powered by RES, are all things that may be used as factors and variables in an optimisation process for the system. The technique known as thermodynamic optimisation often entails doing an economic and thermodynamic study of the system in question, considering energy transfer, exergy destruction, and the cost of producing freshwater. The authors in [54] investigated the exergy analysis of a California RO desalination plant with a capacity of 725 m^3 d^{-1}. After analysing the exergy across the system's primary components of the system, the authors calculated the overall exergy loss caused by the complete system and compared it with the exergy loss caused by an ideal replacement unit. The exergy study of the current plant revealed that the membrane modules were responsible for the destruction of a greater proportion of the entire exergy input than the mixing chamber (0.67% of the total exergy input). The second law of efficiency of the plant was determined to be 4.3%, while the second law of efficiency of the alternative unit was enhanced to 4.9% by including a pressure exchanger with two throttling valves in the brine stream. On the contrary, the second law of efficiency of the plant was calculated to be 4.3%. This improved the system by reducing the amount of pump energy required for the input of saltwater input, resulting in savings of 19.8 kW of electrical power. The impact of irreversibility and cost on solar ORC-RO desalination was investigated using a variety of energy recovery designs, including a pressure exchanger, Pelton wheel turbine, and basic. In conjunction with the Shann El-Shiekh RO unit, these energy recovery devices demonstrate that a pressure exchanger is a more cost-effective alternative to the basic and Pelton wheel turbine versions, while also causing the least amount of exergy loss. In the study referred to in [55], many modifications of the non-dominated classification genetic algorithm (NSGA-II, NSGA-II jumping gene [NSGA-II-JG], and NSGA-II-adapted jumping gene [NSGA-II-aJG]) were used to locate an optimum solution to the problem of multi-objective optimisation of RO desalination units. The target functions were to minimise system cost, permeate throughput, and permeate concentration using choice factors that included alternative operating pressures, active areas of membranes, types of membranes, and module configurations. Compared with the NSGA-II and NSGA-II-JG, the NSGA-II-aJG converges at a faster rate. The technique to thermodynamic optimisation is broken down into its essential parts and summarised in table 4.4.

Table 4.3. Summary of operational optimisation of RES desalination systems [24]. Reprinted from [24], copyright (2021) with permission from Elsevier.

Objective function	Constraints	Optimisation techniques	Outlook	References
Evaluation of economic analysis of the system using a levelized cost matrix and determination of the energy requirement	Physical properties of the system components and year of consideration.	HOMER and ROSA	Seven hybrid RES configurations were evaluated; wind-PV-diesel was considered a better option.	[29]
Evaluation of the LCOE and LCOW, determination of energy requirement and emission reduction	Physical properties of the system components and year of consideration.	ROSA	Significant reduction in emissions.	[33]
Minimisation of critical exceed of electricity production (CEEP) and the highest possible share of RES in the mix	Inequality constraints: storage capacity and allowable RES. Equality constraints: power balance equation.	An optimisation algorithm was developed to select the best combination of hybrid RES.	Increasing the optimisation horizon from 1 to 24 h can reduce the CEEP by 80%.	[49]
Determination of the feasibility of supplying both energy and water from RES at a minimal cost	Power balance equations.	HOMER and ROSA	The technical feasibility and reliability of the system were confirmed and CO_2 was reduced.	[47]
Minimisation of the total cost of the system (cost of power generation + cost of desalination)	Inequality constraints: power and desalination limits. Equality constraints: power balance equation.	Proposed MILP	Significant economic savings. The flexibility of the desalination unit can be advanced by adding demand response (DR) programs, making room for frequency regulation or spinning reserve products.	[50]
Minimisation of energy cost in water supply network	Inequality constraints: pressure head, water tank capacity limits. Equality constraints: water flow balance equation and mass conservation equation.	Formulated an optimal water flow model	Adaptation of the proposed optimisation model and integration with metaheuristic techniques are recommended for implementation in the water network with uncertainty of demand, considering DRs.	[51, 52]

Minimisation of the total annualised production cost while maximising the percentage of RES in the total power production and minimising wind power curtailed	Inequality constraints: limits of the capacity of the water reservoir. Equality constraints: equations for water flow balance and power balance.	Direct multi-search	Increased penetration of RES and reduction of emission at a minimal cost.	[31]
Minimisation of unit operation cost	Technical limits of components, spinning reserve requirements, and efficiency curves at part load operation.	Formulated MILP	Fast and accurate solution to the MILP problem.	[48]
Minimisation of freshwater cost of hybrid MED and RO system	Operational parameters of RO unit, seawater salinity of MED and freshwater salinity of RO-MED.	Sensitivity analysis	The optimal feed pressure and the feed flow rate of RO and stem flow rate along the temperature of MED were regarded continuous variables of optimisation. RO-MED proved to be cost-effective.	[28]
Minimisation of specific energy consumption, total dissolved solids of permeate, and maximisation of permeate production rate	Limits of transmembrane pressure and total dissolved solids of permeate.	GA	The multi-objective optimisation shows the practical benefit of hybrid RES with a jointly controlled module.	[39]
Maximisation of the permeate production rate	Limits of transmembrane pressure and total dissolved solids of permeate.	GA	The two-dimensional GA produced the maximum permeate rate even as the proportional integral derivative failed.	[53]

Table 4.4. Summary of the thermodynamic optimisation of RES desalination processes [24]. Reprinted from [24], copyright (2021) with permission from Elsevier.

Ref.	Objective function	Constraints	Optimisation techniques	Outlook
[55]	Minimisation of the cost of desalination and permeate concentration.	Inequality constraints: membrane permeability and permeate concentration bounds. Equality constraints: component properties and mass conservation equation.	NSGA-II, NSGA-II-JG, NSGA-II-aJG	At the highest permeability of water and salt, maximum permeate throughput is achieved with a constrained permeate concentration. The NSGA-II-aJG converges more rapidly than NSGA-II and NSGA-II-JG.
[56]	Maximisation of energy efficiency, exergy efficiency, desalination output, VAR (vapour absorption refrigeration system) cooling output, and total output.	Inequality constraints: turbine pressure and steam extraction limit and desalination heated feedwater temperature. Equality constraints: energy and mass balance equations.	GA	The polygeneration system is cost-effective and is efficient and considered a very good option to satisfy energy demand in the global energy transition.
[57]	Maximisation of net output power and freshwater from the system using an energy recovery system for wastewater and heat.	Inequality constraints: turbine pressure and steam extraction limit, boiler outlet temperature limit, and heat exchanger temperature difference. Equality constraints: energy and mass balance equations.	GA	Heat recovery from the boiler blowdown stream increased the power production of a steam turbine. The water recovery unit increased freshwater production. Exergy analysis shows high exergy destruction in the boiler unit, and hence requires improvement. Economic analysis indicates an adequate payback time. The optimisation process improves the overall performance of the system by improving the net power output.
[58]	Maximisation of the permeate volumetric flow rate.	Permeate concentration limits and allowable pressure across the membrane.	Genetic algorithm reverse osmosis (GARO)	GARO shows approximate linearity between the permeate volumetric flow rate and the pressure difference across the membrane. Permeate concentration decreased as the permeate volumetric flow rate and pressure difference across the membrane increased.

4.3 Research output and trend

The research output and trend in optimising RES-powered desalination were evaluated using VOSviewer software, which was applied to data acquired from the Clarivate Web of Science database. This database was selected for data extraction in consideration of it being the oldest database and that it has a significant emphasis on research based on scientific principles [59]. Additionally, it has been discovered that there is a considerable amount of overlap between the Web of Science and Scopus (another similarly popular scientific database) in the area of study; hence, it is important to choose between the two to proceed. When researching, the terms 'optimization,' 'renewable-energy sources,' and 'desalination' were utilised as search terms. The years 2001 through 2023 were included in the scope of the search, and the only content types considered were previously published journal articles and proceedings. The search of the relevant literature turned up 128 articles that had a combined total of 765 keywords. A total of 60 keywords were generated for publications with five co-occurrences of the selected keywords. The network visualisation map that indicates the relationship between keywords and frequent terms at five co-occurrences is shown in figure 4.2 while figure 4.3 depicts overlay visualisation.

The network visualisation map forms four clusters differentiated with red, blue, green, and yellow colour, which suggest significant links that determine the focus of the selected articles. Subsequently, the groups are subtitled (i) 'economic feasibility

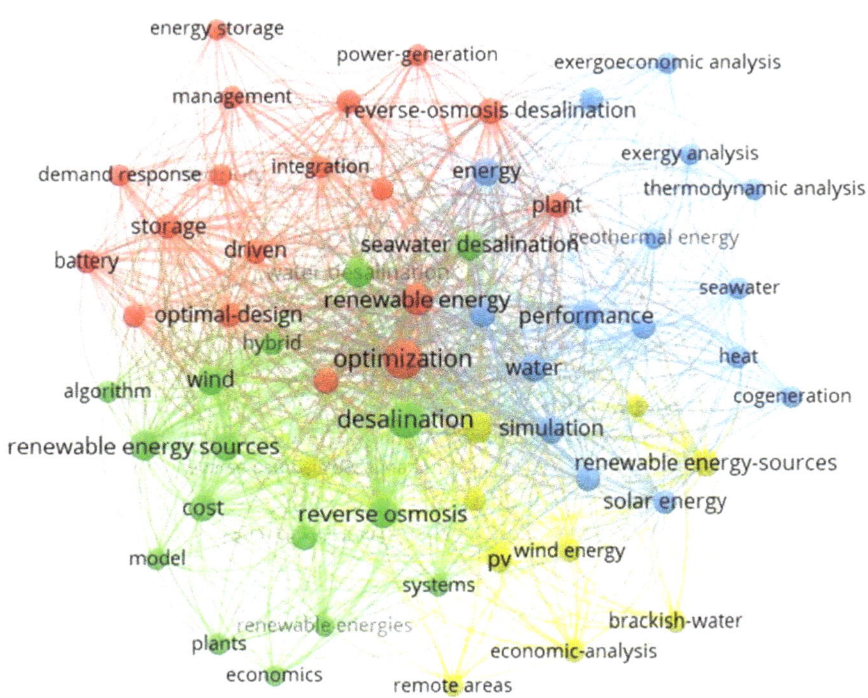

Figure 4.2. Network visualisation map.

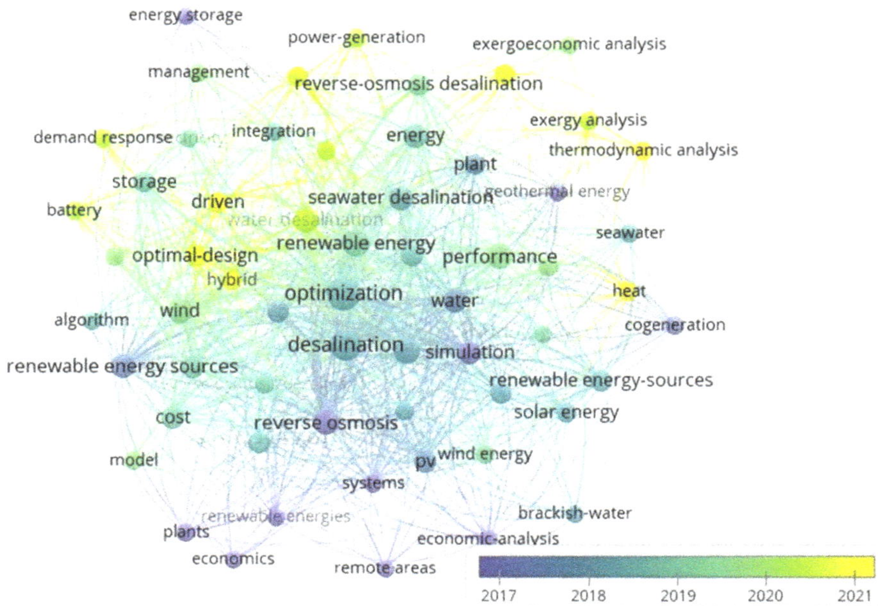

Figure 4.3. Overlay visualisation map of optimisation of RES desalination.

of a renewable energy source—integrated desalination system,' (ii) 'modelling and simulation of RES-powered RO desalination system,' (iii) 'design and exergy analysis of a solar powered desalination unit,' and (iv) optimisation of PVs-RO desalination technology,' respectively.

4.3.1 Economic feasibility of renewable energy source—integrated desalination system

Cost, desalination, economics, feasibility, integration, power generation, renewable energy, RES, solar, storage, system, and wind are among the 18 components of this cluster (red). The strongest correlation is between desalination and renewable energy. Other item links in the cluster suggest that the economic development of an RES desalination system is the research focus of publications in this group, as described in [28, 44, 60, 61]. The emphasis is on the economic and social dimensions of desalination, exploring its cost effectiveness, affordability, and impact on local communities. The section further assesses the socio-economic benefits and challenges associated with implementing large-scale desalination projects, including job creation, regional development, and water equity.

4.3.2 Modelling and simulation of renewable energy source—power RO desalination system

Plant, electricity, RES, RO, seawater desalination, simulation, systems, waste heat, and water desalination are the 16 elements in the second cluster (green). RES and water desalination are the clearest links in this group. Considering the cluster's pattern with other common items, their links suggest that the cluster's research of the group is on modelling and simulation of the integration of RES with seawater RO desalination method, as mentioned in [62, 63]. This shows that modelling and simulation have emerged as indispensable tools to address complex challenges in the fields of energy and water. As demands for sustainable resource management intensify, there is a growing need to understand the intricate interactions and interdependencies between energy systems and water resources. In this context, modelling and simulation offer valuable insight into the dynamics of these critical domains, enabling researchers and policymakers to design informed strategies for a sustainable future.

4.3.3 Design and exergy analysis of solar-powered desalination unit

The third cluster (blue) consists of the following six components: energy, exergy analysis, performance, solar, desalination, and unit. The strongest relationships are between design, energy, and water; as a result, the primary emphasis of the studies in this group is on the performance and energy analyses of the solar-powered desalination technology designed as given by [64, 65]. Exergy analysis assesses the thermodynamic efficiency of the solar-powered desalination unit by evaluating the irreversibilities and losses within the system. It provides valuable insight into the locations of inefficiencies and opportunities for improvement. This study identifies exergy destruction points within the system, highlighting areas that may require optimisation to enhance overall performance. Furthermore, exergy analysis goes beyond traditional energy-based assessments, providing a more comprehensive evaluation of the unit's sustainability, resource utilisation, and environmental impact.

4.3.4 Optimisation of PV-RO technology

This cluster (yellow) consists of 10 different aspects: economic analysis, PVs, remote area, renewable energies, RO, solar energy, and technologies. Since PVs and RO have the strongest connection between them, the primary focus of the research and publications in this group is on the economic analysis of PV-RO in remote areas, as stated by [66, 67]. Among other RES and desalination technologies, PV-RO desalination has emerged as a promising and sustainable solution, utilising the abundance of solar energy and the efficiency of RO to produce freshwater from seawater or brackish sources. Research in this area aims to conduct an economic analysis of the PV-RO desalination system to assess its viability, cost effectiveness, and potential as a long-term sustainable freshwater production method [46, 68]. These studies aim to explore various optimisation strategies to enhance the water production capacity and energy efficiency of PV-RO desalination plants. By maximising freshwater output while minimising energy consumption, these optimised systems hold the potential to meet growing water demands in a cost-effective and environmentally friendly manner. Factors such as panel orientation, array size, and RO system size are analysed to maximise energy capture and freshwater production.

Efficient energy storage and management systems are explored to ensure a stable and continuous freshwater production process, especially during periods of low solar irradiance. The implementation of advanced control algorithms and real-time monitoring systems is examined to optimise the plant's overall performance, reduce energy waste, and enhance system reliability. The energy efficiency of the optimised system is analysed, with a focus on the specific energy consumption per cubic metre of freshwater produced, then compared to conventional desalination technologies.

The overlay visualisation map illustrating the trend of research emphasis can be seen in figure 4.3. The study topics that were the subject of publications from 2017 to 2019 are housed in the purple cluster. From 2019 to 2020, the study focused on regions within the green group, including the optimisation of desalination systems. However, from mid-2020 to 2021, the research was shifting to the yellow cluster, which mainly involves performance assessments of RES with energy storage systems to the power desalination unit.

4.4 Summary

Modelling and simulations have become indispensable tools in the study, design, and optimisation of energy and water desalination systems. As the world faces escalating challenges related to freshwater scarcity and the need for sustainable energy solutions, the integration of modelling techniques has proven essential for developing efficient and economically viable desalination processes. This scholarly review examines the fundamental principles of modelling and simulations in the context of energy and water desalination systems. It explores various modelling approaches, their applications, and the insights they offer to improve the performance and sustainability of desalination technologies. It is a comprehensive assessment of the optimisation of the RES-RO desalination system, with a focus on optimum system size, optimal system operation, and optimal thermodynamic analysis. Challenges and advances in different optimisation techniques as applicable to integrated energy–water models were identified. Therefore, the following findings and conclusions are drawn from the study.

The techno-economic analysis of RES-RO desalination has been a prominent topic of study; however, the optimisation of thermodynamic parameters such as exergy destruction and specific energy consumption has not been addressed to a significant extent. As a result, more studies should be done in the future that use optimisation approaches to thermodynamic analysis of the RES-RO desalination system.

A combined optimisation strategy, which takes into account the size, operation, and thermodynamic impact of the system, should be explored for use in the optimisation and analysis of RES-RO systems to achieve a more comprehensive level of results. In addition, there has been a recent tendency toward the use of various metaheuristic optimisation approaches, which has led to a change in focus away from mathematical methodologies. On the other hand, mathematical methods continue to show a great deal of promise for producing correct answers, particularly in light of the many alternative computer software solvers that are now available. There is a need for more investigation into metaheuristic optimisation strategies, and yet, only a relatively small number of these approaches have been used in this field. In addition, a stronger focus on economic performance metrics has been used to direct optimisation efforts in this domain. When creating objective functions in the future, a more thorough study should use both the economic and the reliability index. Integrating AI and machine learning techniques, and cross-disciplinary collaboration is the pathway to advance the state of the art in sustainable desalination technologies. Additionally, big data analytics to enhance the simulation, and cloud computing with distributed simulation will address uncertainties such as climate change and its impact on the energy–water nexus and socio-economic uncertainties and further help in managing stakeholder engagement and feedback. In general, modelling and simulation play a vital role in advancing knowledge, making informed decisions, and mitigating risks in numerous domains. They allow researchers to explore different scenarios without affecting the real system, saving time and resources, and often providing valuable insights that might be challenging to obtain through direct experimentation alone.

References

[1] Ahmed F E, Hashaikeh R, Diabat A and Hilal N 2019 Mathematical and optimisation modelling in desalination: state-of-the-art and future direction *Desalination* **469** 114092

[2] Basnet S, Deschinkel K, Le Moyne L and Cécile Péra M 2023 A review on recent standalone and grid-integrated hybrid renewable energy systems: system optimisation and energy management strategies *Renew. Energy Focus* **46** 103–25

[3] Barton P I and Pantelides C C 1994 Modeling of combined discrete/continuous processes *AIChE J.* **40** 966–79

[4] Pitts J, Gopal S, Ma Y, Koch M, Boumans R M and Kaufman L 2020 Leveraging big data and analytics to improve food, energy and water system sustainability *Front. Big Data* **3** 1–16

[5] Subramanian A S R, Gundersen T and Adams T A 2018 Modeling and simulation of energy systems: a review *Processes* **6** 238

[6] Grossmann I E and Guillén-Gosálbez G 2010 Scope for the application of mathematical programming techniques in the synthesis and planning of sustainable processes *Comput. Chem. Eng.* **34** 1365–76

[7] Brumana G, Franchini G, Ghirardi E and Perdichizzi A 2022 Technoeconomic optimization of hybrid power generation systems: a renewables community case study *Energy* **246** 123427

[8] Wei M, Wen Y, Meng Q, Zheng S, Luo Y and Liao K 2023 Combined hybrid energy storage system and transmission grid model for peak shaving based on time series operation simulation *Glob. Energy Interconnect.* **6** 154–65

[9] Dehghani H and Vahidi B 2022 Evaluating the effects of demand response programmes on distribution cables life expectancy *Electr. Power Syst. Res.* **213** 108710

[10] Schweighofer B, Buchroithner A, Felsberger R and Wegleiter H 2023 Optimized selection of component models for photovoltaic and energy storage system simulations *Sol. Energy* **249** 559–68

[11] Sun X, Lin L, Jin C, Guo Z and You Y 2023 A system dynamics-based lifecycle carbon footprint calculation model for integrated energy stations *Energy Rep.* **9** 1726–33

[12] Sobana S and Panda R C 2011 Review on modelling and control of desalination system using reverse osmosis *Rev. Environ. Sci. Biotechnol.* **10** 139–50

[13] Blanco-Marigorta A M, Lozano-Medina A and Marcos J D 2017 A critical review of definitions for exergetic efficiency in reverse osmosis desalination plants *Energy* **137** 752–60

[14] Salakhi M, Eghtesad A and Afshin H 2022 Heat and mass transfer analysis and optimisation of freeze desalination utilizing cold energy of LNG leaving a power generation cycle *Desalination* **527** 115595

[15] Pérez G, Gómez P, Ortiz I and Urtiaga A 2022 Techno-economic assessment of a membrane-based wastewater reclamation process *Desalination* **522** 115409

[16] Alsadaie S M and Mujtaba I M 2017 Dynamic modelling of heat exchanger fouling in multistage flash (MSF) desalination *Desalination* **409** 47–65

[17] Hills S, Dana S and Wang H 2021 Dynamic modelling and simulation of nuclear hybrid energy systems using freeze desalination and reverse osmosis for clean water production *Energy Convers. Manag.* **247** 114724

[18] Ben Bacha H, Bouzguenda M, Abid M S and Maalej A Y 1999 Modelling and simulation of a water desalination station with solar multiple condensation evaporation cycle technique *Renew. Energy* **18** 349–65

[19] Johannsen R M, Prina M G, Østergaard P A, Mathiesen B V and Sparber W 2023 Municipal energy system modelling—a practical comparison of optimisation and simulation approaches *Energy* **269** 126803

[20] Arora M, Nair S, Malano H and George B 2020 A combined simulation-optimisation modelling framework for assessing the energy use of urban water systems *J. Environ. Manage.* **274** 111166

[21] Wang C, Chen Y, Sun M and Wu J 2023 Potential of technological innovation to reduce the carbon footprint of urban facility agriculture: a food–energy–water–waste nexus perspective *J. Environ. Manage.* **339** 117806

[22] Huang J, Tan Q, Zhang T and Wang S 2023 Energy–water nexus in low carbon electric power systems: a simulation-based inexact optimisation model *J. Environ. Manage.* **338** 117744

[23] Rao S S 2019 *Engineering Optimization Theory and Practice* (New York: Wiley)

[24] Okampo E J and Nwulu N 2021 Optimisation of renewable energy powered reverse osmosis desalination systems: a state-of-the-art review *Renew. Sustain. Energy Rev.* **140** 110712

[25] Twaha S and Ramli M A M 2018 A review of optimization approaches for hybrid distributed energy generation systems: off-grid and grid-connected systems *Sustain. Cities Soc.* **41** 320–31

[26] Ooka R and Ikeda S 2015 A review on optimization techniques for active thermal energy storage control *Energy Build.* **106** 225–33

[27] Acuña L G, Lake M, Padilla R V, Lim Y Y, Ponzón E G and Soo Too Y C 2018 Modelling autonomous hybrid photovoltaic wind energy systems under a new reliability approach *Energy Convers. Manag.* **172** 357–69

[28] Al-Obaidi M A, Filippini G, Manenti F and Mujtaba I M 2019 Cost evaluation and optimisation of hybrid multi-effect distillation and reverse osmosis system for seawater desalination *Desalination* **456** 136–49

[29] Gökçek M 2018 Integration of hybrid power (wind-photovoltaic-diesel-battery) and seawater reverse osmosis systems for small-scale desalination applications *Desalination* **435** 210–20

[30] Koutroulis E and Kolokotsa D 2010 Design optimization of desalination systems power-supplied by PV and W/G energy sources *Desalination* **258** 171–81
[31] Dui N, Carvalho M G, Madeira J F A, Segurado R and Costa M 2016 Optimization of a wind powered desalination and pumped hydro storage system *Appl. Energy* **177** 487–99
[32] Zejli D, Benchrifa R, Bennouna A and Zazi K 2004 Economic analysis of wind-powered desalination in the south of Morocco *Desalination* **165** 219–30
[33] Gökçek M and Gökçek Ö B 2016 Technical and economic evaluation of freshwater production from a wind-powered small-scale seawater reverse osmosis system (WP-SWRO) *Desalination* **381** 47–57
[34] Maleki A, Khajeh M G and Rosen M A 2016 Weather forecasting for optimisation of a hybrid solar-wind–powered reverse osmosis water desalination system using a novel optimiser approach *Energy* **114** 1120–34
[35] Zhang G, Wu B, Maleki A and Zhang W 2018 Simulated annealing-chaotic search algorithm-based optimization of reverse osmosis hybrid desalination system driven by wind and solar energies *Sol. Energy* **173** 964–75
[36] Maleki A 2018 Design and optimization of autonomous solar-wind-reverse osmosis desalination systems coupling battery and hydrogen energy storage by an improved bee algorithm *Desalination* **435** 221–34
[37] Molinos-Senante M and González D 2018 Evaluation of the economics of desalination by integrating greenhouse gas emission costs: an empirical application for Chile *Renew. Energy* **133** 1327–37
[38] Karabelas A J, Koutsou C P, Kostoglou M and Sioutopoulos D C 2018 Analysis of specific energy consumption in reverse osmosis desalination processes *Desalination* **431** 15–21
[39] Lee S, Cho H Y and Har D 2018 Operation optimization with jointly controlled modules powered by hybrid energy source: a case study of desalination *Renew. Sustain. Energy Rev.* **81** 3070–80
[40] Abdelshafy A M, Hassan H and Jurasz J 2018 Optimal design of a grid-connected desalination plant powered by renewable energy resources using a hybrid PSO–GWO approach *Energy Convers. Manag.* **173** 331–47
[41] Maleki A, Pourfayaz F and Ahmadi M H 2016 Design of a cost-effective wind/photovoltaic/hydrogen energy system for supplying a desalination unit by a heuristic approach *Sol. Energy* **139** 666–75
[42] Abdelshafy A M, Hassan H and Jurasz J 2018 Optimal design of a grid-connected desalination plant powered by renewable energy resources using a hybrid PSO-GWO approach *Energy Convers. Manag.* **173** 331–47
[43] Smaoui M, Abdelkafi A and Krichen L 2015 Optimal sizing of stand-alone photovoltaic/wind/hydrogen hybrid system supplying a desalination unit *Sol. Energy* **120** 263–76
[44] Hossam-Eldin A, El-Nashar A M and Ismaiel A 2012 Investigation into economical desalination using optimized hybrid renewable energy system *Int. J. Electr. Power Energy Syst.* **43** 1393–400
[45] Okampo E J and Nwulu N I 2020 Optimal energy mix for a reverse osmosis desalination unit considering demand response osmosis *J. Eng. Des. Technol.* **18** 1287–303
[46] Bourouni K, Ben M'Barek T and Al Taee A 2011 Design and optimisation of desalination reverse osmosis plants driven by renewable energy using genetic algorithms *Renew. Energy* **36** 936–50

[47] Prieto-Prado I, Del Ro-Gamero B, Gómez-Gotor A and Pérez-Báez S O 2018 Water and energy self-supply in isolated areas through renewable energies using hydrogen and water as a double storage system *Desalination* **430** 1–14
[48] Astolfi M, Mazzola S, Silva P and Macchi E 2017 A synergic integration of desalination and solar energy systems in stand-alone microgrids *Desalination* **419** 169–80
[49] Duić N *et al* 2015 Modeling of optimal energy flows for systems with close integration of seawater desalination and renewable energy sources: case study for Jordan *Energy Convers. Manag.* **110** 249–59
[50] Mohammadi F, Sahraei-Ardakani M and Al-Abdullah Y 2018 Coordinated operation of power generation and water desalination *IFAC-PapersOnLine* **51** 456–61
[51] Fooladivanda D and Taylor J A 2018 Energy-optimal pump scheduling and water flow *IEEE Trans. Control Netw. Syst.* **5** 1016–26
[52] Okampo E J and Nwulu N 2021 Techno-economic evaluation of reverse osmosis desalination system considering emission cost and demand response *Sustain. Energy Technol. Assess.* **46** 101252
[53] Lee S, Myung S, Hong J and Har D 2016 Reverse osmosis desalination process optimised for maximum permeate production with renewable energy *Desalination* **398** 133–43
[54] Cerci Y 2002 Exergy analysis of a reverse osmosis desalination plant in California *Desalination* **142** 257–66
[55] Guria C, Bhattacharya P K and Gupta S K 2005 Multi-objective optimization of reverse osmosis desalination units using different adaptations of the non-dominated sorting genetic algorithm (NSGA) *Comput. Chem. Eng.* **29** 1977–95
[56] Sahoo U, Kumar R, Singh S K and Tripathi A K 2018 Energy, exergy, economic analysis and optimisation of polygeneration hybrid solar-biomass system *Appl. Therm. Eng.* **145** 685–92
[57] Noroozian A, Mohammadi A, Bidi M and Ahmadi M H 2017 Energy, exergy and economic analyses of a novel system to recover waste heat and water in steam power plants *Energy Convers. Manag.* **144** 351–60
[58] Djebedjian B, Gad H, Khaled I and Rayan M A 2008 Optimization of reverse osmosis desalination system using genetic algorithms technique *Twelfth International Water Technology Conference* 1047–67
[59] Aghaei Chadegani A *et al* 2013 A comparison between two main academic literature collections: Web of Science and Scopus databases *Asian Soc. Sci.* **9** 18–26
[60] Díez E, Langston P, Ovejero G and Romero M D 2009 Economic feasibility of heat pumps in distillation to reduce energy use *Appl. Therm. Eng.* **29** 1216–23
[61] Baldea M *et al* 2007 Feasibility study on wind-powered desalination *Desalination* **203** 463–70
[62] Kelley L C and Dubowsky S 2013 Thermal control to maximise photovoltaic powered reverse osmosis desalination systems productivity *Desalination* **314** 10–9
[63] He W, Wang Y and Shaheed M H 2015 Stand-alone seawater RO (reverse osmosis) desalination powered by PV (photovoltaic) and PRO (pressure retarded osmosis) *Energy* **86** 423–35
[64] Nafey A S, Sharaf M A and García-Rodríguez L 2010 Thermo-economic analysis of a combined solar organic Rankine cycle reverse osmosis desalination process with different energy recovery configurations *Desalination* **261** 138–47
[65] Gude V 2018 Exergy evaluation of desalination processes *ChemEngineering* **2** 28

[66] Abdallah S, Abu-Hilal M and Mohsen M S 2005 Performance of a photovoltaic powered reverse osmosis system under local climatic conditions *Desalination* **183** 95–104
[67] Garg M C and Joshi H 2014 Optimization and economic analysis of small-scale nanofiltration and reverse osmosis brackish water system powered by photovoltaics *Desalination* **353** 57–74
[68] Poullikkas A 2010 An optimization model for the production of desalinated water using photovoltaic systems *Desalination* **258** 100–5
[69] Okampo E J, Nwulu N and Bokoro P N 2022 Optimal placement and operation of FACTS technologies in a cyber-physical power system: critical review and future outlook *Sustainability* **14** 7707

Part IV

Techno-economic and environmental consideration

IOP Publishing

Optimisation of Renewable Energy Powered Desalination Systems
A sustainable techno-economic and environmental consideration
Ewaoche John Okampo, Nnamdi Nwulu and Pitshou Ntambu Bokoro

Chapter 5

Demand-side management: a tool for energy–water efficient management

In an era of growing energy demand, water scarcity, and environmental concerns, the imperative for efficient resource management has become paramount. This chapter delves into the concept of demand-side management (DSM) as a powerful tool for achieving energy–water efficiency and sustainability. DSM involves a set of strategies and practices aimed at optimising energy consumption patterns on the consumer end. The chapter explores how DSM extends its influence beyond energy to encompass water resources, creating a comprehensive framework for managing the intertwined energy–water nexus. Moreover, DSM's integration with advanced technologies and real-time data analytics empowers consumers to make informed decisions about their energy and water usage, fostering a culture of resource consciousness. It highlights how load shifting, demand response programmes, and smart grid technologies can be tailored to address the unique challenges of energy–water management, enhancing system resilience, and reducing environmental impact. Furthermore, the chapter emphasises the importance of cross-sectoral collaboration among utilities, regulators, industries, and consumers to harness the full potential of DSM for energy–water efficient management. In conclusion, this chapter underscores DSM's pivotal role as a versatile and effective tool for achieving energy–water efficient management. By aligning consumption patterns, optimising infrastructure, and promoting conscious resource use, DSM offers a pathway toward a more sustainable future where energy and water resources are conserved, harnessed efficiently, and harmonised for the well-being of both present and future generations.

doi:10.1088/978-0-7503-6154-5ch5

5.1 Introduction

The fundamental nature of energy and water in modern civilisation cannot be overstated. These two essential resources drive the core processes underpinning society's growth and survival, as [1] noted. Their role in our daily lives is profound and far-reaching. From illuminating our homes at night to fuelling the machines in vast factories, energy is a driving force. On the other hand, water is crucial for our food supply, not just for consumption but also in the agricultural sector, which relies heavily on it. According to [2], agriculture alone accounts for an astonishing 70% of all freshwater withdrawals across the globe. This immense demand for energy and water is not a recent phenomenon. With the advent of industrialisation, society's thirst for these resources has increased exponentially. They have yet to operate in isolation either. There exists a deeply intricate relationship between energy and water, which experts often term the energy–water nexus. As elucidated by [3], the utilisation or production of one has inevitable ramifications on the other. As technology and industries grew, especially in the 20th century, the intertwined nature of this relationship became even more evident. A classic example is that of thermal power plants that rely heavily on water for cooling purposes. In a similar vein, hydraulic fracturing, a method used to extract underground resources, requires copious amounts of water. On the flip side, processes that ensure a steady supply of water, such as treatment or distribution, invariably consume a considerable amount of energy. Regions characterised by significant industrial activities or those dealing with water scarcity exemplify this interconnected relationship.

Several socio-economic factors have compounded the challenges associated with managing this nexus. Population spikes, rapid urban development, and the over-arching theme of economic expansion have exerted immense pressure on these resources, as [4] pointed out. In the context of globalisation and the surge of urban migration, countries that are rapidly urbanising, such as India and China, serve as focal points for this accelerated demand, as observed by [5]. Looking at the history for solutions, the immediate response to address this spiralling demand had been predominantly on the supply side. The conventional wisdom suggested that building more power plants or identifying and exploiting new water sources would solve the quandary. However, as the latter half of the 20th century unfolded, the pitfalls of such a supply-centric strategy became painfully apparent. The environmental, social, and economic costs began to mount. The Three Gorges Dam, located in China, and the extensive environmental and social consequences of coal-fired power plants serve as stark reminders of the limitations of solely relying on increasing supply, as referenced by [6].

It is with the backdrop of these challenges that the idea of DSM emerged as a potential panacea. Gaining momentum primarily in the last few decades of the 20th century, DSM presented a fresh perspective. Instead of focusing on increasing the supply, it proposed managing and even reducing resource consumption at the end-user level. The adoption of energy-efficient devices and fixtures that conserve water can be credited to the ethos of DSM. The results were palpable, with evident resource conservation, resulting in both economic and environmental dividends.

Yet, despite the promise shown by DSM, there is still an evident void in research. Although the potential of DSM is undeniable, comprehensive studies examining its relevance in the context of the energy–water nexus are conspicuously absent. Much of the existing academic literature and research seem to tilt towards either energy or water. This oversight is not just an academic concern. In terms of practical application and the crafting informed policies, understanding this nexus and the role of DSM within it is imperative. As [7] succinctly put it, the judicious management of energy and water has far-reaching implications spanning health, food security, and economic growth.

In light of these challenges and the existing knowledge gap, there is a pressing need for research that delves deeper into the application of DSM within the context of the energy–water nexus. This research, as conceptualised, seeks to be a beacon in this direction. By amalgamating data from various sources, leveraging cutting-edge modelling techniques, and undertaking cross-sectoral analysis, the objective is clear: to carve a roadmap that can guide stakeholders, inform policymakers, and aid fellow researchers. As highlighted by [8], such an endeavour has the potential to set society on a path to a sustainable future, one where energy and water are managed efficiently and judiciously.

5.2 Understanding DSM

5.2.1 Definition and concepts of DSM

DSM, as it is commonly referred to, represents a significant paradigm shift in the approach of energy utility companies towards their consumers. As [9] lucidly explains, DSM is not just another programme but a robust strategy aimed at motivating end-users—consumers and businesses alike—to bring about substantive changes in both the magnitude and the patterns of their electricity usage. One could say that it is a nudge, a gentle push for the consumers to rethink their consumption habits and not just how much they consume but also when they consume. The timing of electricity consumption, [10] argues, is as crucial as the volume itself. DSM, in its intricate design, seeks to achieve a dual goal. On the one hand, it aims to curtail the peak demand for electricity. Why is this significant? High peak demands can strain the electricity grid enormously, leading to potential blackouts and a pressing need to establish new generation capacities. This is not merely about preventing short-term outages; it has long-term implications on investment in infrastructure, environmental considerations, and costs that may get passed on to the end consumers. On the other hand, DSM plays a pivotal role in the grand orchestration of electricity grids. It aids in striking a delicate balance, an equilibrium, if you will, between the demand and supply of electricity.

However, the utility of DSM is not confined to just regulating the grid or curtailing consumption. The authors of [11] further emphasise that DSM has a broader purpose. Beyond the obvious goal of reducing energy consumption across the board, DSM has another equally pertinent objective. It aims to shift or transfer the demand for electricity from peak hours to off-peak hours. This not only lightens the load on the grid during high-demand periods but also ensures that energy systems operate optimally, maximising efficiency across the entirety of the grid. One of the cornerstones of DSM is energy efficiency. The authors of [12] delve into the semantics of energy efficiency within the DSM landscape, describing it as an overarching strategy rooted in a simple yet profound objective: to diminish the overall amount of energy required to achieve specific tasks or outcomes. Achieving this is no small feat. It demands a multi-pronged approach. This involves the integration of technologies that are inherently designed to be conservative in their energy consumption. But the journey does not end with just introducing energy-saving technologies. It extends to the promotion of behavioural changes among consumers. People must be made aware, educated, and often incentivised to adopt practices that lead to efficient energy use. Additionally, processes, whether industrial, commercial, or residential, can be refined and enhanced to further the cause of energy efficiency.

In the vast realm of DSM, another crucial concept emerges—demand response (DR). The authors of [13] elucidate the nuance of DR, which can be visualised as the reflexive reaction of the electricity grid. It is predicated on altering an electricity consumer's consumption based on the ebbs and flows of electricity pricing [14]. The mechanism is quite ingenious. When electricity prices soar or the grid's stability is precarious, consumers are incentivised to cut down on their consumption. It is a responsive action, a countermeasure to ensure that the grid remains stable and electricity remains available to all without disruptions [15]. But DSM does not operate solely on energy efficiency and DR principles. Enter load management, another integral facet that [16] emphasises. Think of load management as an orchestra conductor, ensuring that every instrument (in this case, electricity supply sources) plays in harmony. Load management directs the electricity consumption in such a way that it aligns closely with the available supply, ensuring seamless operation. This involves a gamut of strategies, each distinct but interrelated. 'Peak clipping' is one strategy that aims to truncate the demand during peak consumption hours. Its counterpart, 'valley filling,' does the opposite: it encourages increased demand during off-peak times. There are strategies like load shifting, strategic conservation, and flexible load shape. All these diverse methods and techniques converge toward a singular goal—to ensure that the electricity grid functions reliably and efficiently.

The march of technology has left no stone unturned, and DSM is no exception. The world is witnessing a transformative phase in DSM, primarily powered by technological innovations such as smart grids and smart metering. The authors of [17] draw attention to these revolutionary tools. Outfitted with the latest technological prowess, they provide consumers with real-time data on their energy consumption patterns and associated costs. The implications are profound. Armed with these data, consumers are in a position of power. Now they can accurately monitor, analyse, and adjust their consumption patterns. Such tools make users more aware and greatly enhance the effectiveness of the strategies falling under DSM, pushing the boundaries of what is possible in the quest for a more sustainable and efficient energy future.

The concept of DR is increasingly being explored and adapted for use in water systems. Its application in water systems involves modifying water consumption patterns to match supply availability or optimise water resource management. This approach aims to improve water efficiency, reduce waste, and mitigate the impact of water scarcity and variability. It has been applied in smart irrigation systems, industrial water management, water reuse, and desalination. The application of DR principles in water systems offers an innovative approach to optimise water management, enhance water efficiency, and mitigate the impact of water scarcity. By encouraging consumers, industries, and utilities to adjust their water consumption patterns based on real-time data and demand conditions, the response to demand contributes to a more sustainable and resilient future of water usage [18].

Considering the importance of energy and water resources and the benefits of their integrated system, DR applications in the energy–water nexus promise efficient utilisation of the resources and sustainability of the system [19]. It involves the use of demand management strategies to optimise energy and water usage. Integrating DR in energy–water systems offers a holistic approach to address the interdependencies between these critical resources, enhancing their sustainability and efficiency [20]. The subsequent sections of this chapter present types of DR programmes, some benefits, and key DR applications in energy–water systems.

5.2.2 Types of DR programmes

The DR programme types can be classified as time-based or incentive-based, as illustrated in figure 5.1. The time-based DR programme involves varying the price in response to load demand, while the incentive-based DR programme offers incentives or penalties to consumers based on them and the service provider [21].

The time-based DR programme could be the time-of-use (ToU) DR programme. It offers a regulated pricing rated on the time of the use of resources (electricity or water). For example, electricity rates are higher during peak hours and lower during off-peak hours. Consumers who shift energy-intensive activities to off-peak periods take advantage of lower rates. This reduces the cost for consumers and the share load demand over time. Thus, we balance demand and supply in an efficient and cost-effective manner. Electricity prices rise significantly during critical peak hours when demand is exceptionally high [23]. During this time, the electricity provider may resort to peak-price critical pricing, with higher prices to discourage consumers

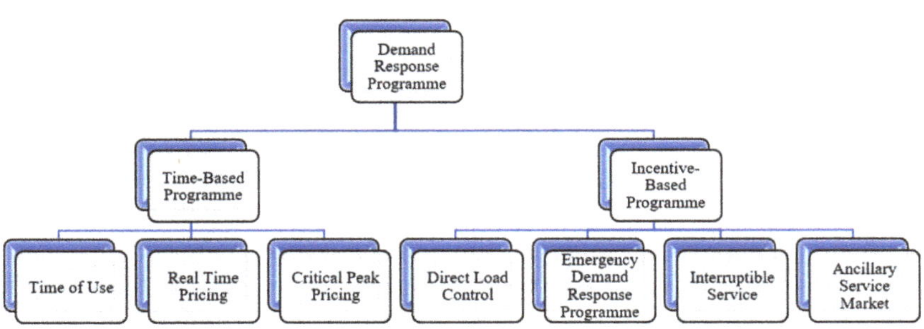

Figure 5.1. Types of DR programmes. Reprinted from [22], copyright (2023) with permission from Elsevier.

at this time. Real-time pricing involves tariff variation per hour. Therefore, adaptability is difficult for consumers [22].

The incentive-based DR programme operates using incentives and penalties during periods of peak demand to reduce the overall load. It includes direct load control (DLC), emergency demand response programme (EDRP), curtailable interruptible (I/C), capacity market (CAP), and ancillary services (A/S). DLC is a sponsored DR programme that involves the direct on and off of high-energy appliances in an attempt to reduce load during the peak period [24, 25]. EDRP is deployed to manage and restore grid stability in an emergency load distortion period [26]. The DLC and EDRP are voluntary initiatives that do not impose fines on consumers who choose not to reduce their energy use, but the I/C and CAP programmes are mandatory programmes. Consumers who fail to curtail their load as advised will face penalties. A/S provide consumers with the possibility to reduce their power consumption in the electrical market, serving as an operating reserve [22].

An effective DR programme requires flexibility, real-time responsiveness, and price signalling. DR is based on the flexibility of electricity consumers to modify their energy consumption patterns when requested or incentivised by the grid operator. Consumers can shift their usage to off-peak hours or reduce nonessential loads during peak periods. Successful DR programmes require timely and real-time communication between the grid operator and consumers [27]. Advanced technologies, such as smart metres and home automation systems, enable real-time monitoring and control of energy consumption [28]. Time-based pricing or critical peak pricing is a common strategy to encourage DR. Under this approach, electricity prices vary depending on demand and supply conditions, providing consumers with price signals to adjust their consumption behaviour.

5.3 Importance and benefits of DSM in energy and water systems

The move towards a more sustainable energy future has witnessed the rise of DSM, heralding the dawn of a new era in energy systems. The authors of [1] underscore the significance of DSM, highlighting its key role in the evolution towards decentralised energy mechanisms. At the heart of this shift lies the pressing imperative to integrate locally sourced renewable energy into the matrix of energy consumption. The implications of such a shift are profound, not just in the manner energy is consumed, but also in how it shapes the broader landscape of energy distribution and sustainability.

DSM is not a mere tactical instrument; it envisions a broader scope. The authors of [2] amplify the essence of DSM by explaining its dual objective: firstly, to augment energy efficiency, and secondly, to reduce overall energy demand. By achieving these twin goals, energy systems are poised to become more sustainable and economically streamlined. One can imagine the transformative impact of an energy ecosystem that is simultaneously environmentally friendly and financially efficient.

DSM's importance does not stop at managing demand or improving efficiency. Its role in grid stability is paramount, particularly in the contemporary world where a consistent power supply is the cornerstone of modern living. According to [8], the predictable demand curve, a product of effective DSM, reinforces the resilience of electrical grids. Such predictability significantly reduces the peril of blackouts, especially during moments of increased energy demand.

The discourse on energy efficiency and grid stability naturally leads to renewable energy. The authors of [6] focus on one of the most salient features of DSM: its ability to weave RES into the grid seamlessly. They emphasise that DSM is adept at synchronising energy demand with periods of prolific renewable energy output, such as during peak sunlight for solar power or optimal wind conditions for wind energy. The emergence and integration of distributed energy resources (DERs) further illustrates DSM's critical role. These resources, including solar panels and wind turbines, are gradually becoming fixtures in modern energy landscapes. Yet, as [11] contends, the true potential of these DERs is realised only when energy demand is managed effectively, a task that DSM is explicitly designed to undertake. By optimising the use of DERs, DSM ensures that these resources are not only effective, but also feasible for broader adoption.

In the grand narrative of environmental conservation, DSM becomes an essential pillar that supports the global endeavour to limit greenhouse gas emissions. By promoting energy efficiency and endorsing RES, DSM contributes directly to the broader objectives of climate change mitigation. Beyond environmental considerations, the marketplace's undercurrents are also affected. The authors of [29] postulate that DSM plays an integral role in the transition from traditional energy market constructs to models that are more adaptive and centred around the consumer.

When one delves into the tangible results of DSM, the benefits become palpable. Both consumers and utility providers stand to reap significant financial rewards. As [30] delineates, these monetary gains stem from enhanced energy efficiency and the strategic reduction of energy consumption during high-demand scenarios. The broader ambition of transitioning to a low-carbon economy also finds its momentum through DSM. The authors of [31] articulated that DSM's influence can considerably reduce the reliance on fossil fuels, inevitably leading to a decline in greenhouse gas emissions. Furthermore, [32] illuminates another ancillary advantage: by spreading out energy demand more uniformly, DSM enables optimal utilisation of the current infrastructure. This eliminates the pressing need for additional infrastructure investments or replacing existing setups. On the economic development front, capital influx related to DSM-driven projects often catalyses job creation, particularly in sectors like construction and manufacturing [33].

Though distinct from energy grids, water systems also resonate with the principles of DSM. The intricate interplay between water and energy underscores the essence of DSM in water management. As [34] elucidates, the energy expended in various water-related processes can be considerable. Therefore, rationalising water demand can indirectly lead to significant energy conservation. With rapid urbanisation on the horizon, DSM's role becomes pivotal in fostering sustainable water resource management. Such initiatives pave the way for the development of resilient urban ecosystems, as underscored by [35]. The agricultural sector, known for its high water consumption, also benefits from DSM strategies. These strategies, as highlighted by

[36], optimise irrigation practices, ensuring water is used judiciously. In regions that experience fluctuating water availability across seasons, [37] advocates for DSM as an essential tool to align demand with supply. Beyond mere management, the environmental benefits of DSM in water systems are noteworthy. The authors of [38] argue that effective DSM programmes can potentially enhance water quality, primarily by reducing the strain on water treatment facilities.

The manifold benefits of DSM in water systems are far-reaching. The authors of [39] emphasise that DSM's effective implementation leads to appreciable water conservation, resulting in monetary savings for both consumers and utility providers. Moreover, the environmental dimension is also catered to; reduced water usage invariably means less energy is expended in the treatment and distribution of water. This, as [40] illustrates, contributes to a smaller carbon footprint for the water sector. DSM's role is not confined to just efficiency and conservation. The authors of [41] underscore DSM's contribution to social equity, emphasising that judicious water management ensures equitable distribution, making water accessible to all segments of society. Furthermore, from a public health vantage point, consistent water supply, facilitated by DSM, is vital. This guarantees access to uncontaminated water, which remains indispensable for sanitation and hygiene, a point accentuated by [17]. In conclusion, similar to its impact on energy systems, DSM acts as a catalyst in the water sector, inspiring innovations and birthing novel water conservation techniques and services, a sentiment echoed by [42].

5.3.1 Role of DSM in sustainable resource management

DSM is critical in sustainable resource management. This domain focusses on strategies adopted primarily by utility companies to persuade end-users to alter their energy and resource consumption behaviours. The overarching goal of these strategies is to usher in more efficient energy consumption patterns while emphasising the conservation of resources [1]. In a parallel context, the domain of sustainable resource management can be broadly defined as the judicious utilisation of various resources to ensure that they remain available for the benefit and consumption of future generations [8]. An intriguing synthesis of these two concepts, DSM and sustainable resource management, yields an integrated ecosystem. This ecosystem emphasises the importance of resource conservation and ensures that consumption is optimised.

The manifold programmes under the DSM's banner aim to diminish energy wastage significantly. This is largely accomplished by advocating for and adopting energy-efficient practises and habits [1]. An important illustration of such a practice involves the adoption and widespread use of energy-efficient appliances. Furthermore, infrastructural modifications, such as increasing the insulation of buildings and other structures, serve as a testament to the same objective [6]. Collectively, these strategic measures lead to a tangible decline in overall energy demand. This reduction is fundamental to promoting sustainable consumption since it minimises the imperative for ramping up energy production. However, the role and impact of DSM go far beyond merely ensuring energy conservation. It intrinsically integrates RES into the broader energy spectrum. Some of the quintessential techniques integral to DSM include load shifting and peak shaving. These methodologies guarantee a seamless equilibrium between the concurrent energy supply and demand. This equilibrium becomes even more paramount considering the integration of RES with intermittent outputs, such as solar and wind [43]. The DSM's role continues here. By fostering a balanced energy consumption blueprint, DSM effectively alleviates the stresses exerted on the power grid. Subsequently, this mitigation leads to enhanced grid resilience, an elemental foundation for fostering a robust and sustainable energy infrastructure. Consequently, there are reduced chances of power interruptions, ensuring uninterrupted electricity access for consumers. From an environmental perspective, DSM indirectly facilitates the reduction of carbon emissions [44]. By emphasising energy efficiency and championing the integration of renewable energy, DSM significantly reduces the dependency on fossil fuels. This shift directly results in a marked reduction in greenhouse gas emission.

Switching the perspective from energy to water, it becomes evident that the principles central to DSM can be adapted and implemented to champion water conservation [45]. For example, the advent and integration of water-efficient appliances and innovative water recycling methodologies are a bulwark against water wastage. However, when considering water management, DSM's influence is not restricted to conservation. By effectively monitoring water demand, DSM substantially prolongs the operational life of water infrastructure. This is achieved by reducing the operational stresses these systems typically endure [46]. As a direct benefit, the water systems not only function optimally, but also the instances of expensive system overhauls are drastically reduced. Considering the global challenge of climate change, several regions worldwide are grappling with acute water scarcity [30]. Within this challenging paradigm, DSM provides a ray of hope. Through strategies that encompass water rationing and the identification and utilisation of alternative water sources, DSM offers viable solutions for communities to adapt and thrive. The unchecked and excessive extraction of water resources has cascading negative effects on aquatic ecosystems [47]. However, with DSM's effective water-demand management, there is a renewed emphasis on preserving these fragile ecosystems, ensuring the survival of the rich biodiversity they host.

Delving deeper into the nexus between energy and water systems, there exists an intrinsic and complex relationship, often termed the energy–water nexus [48]. Effective implementation of DSM strategies in one sector often results in tangential benefits in the other. To elucidate, water conservation invariably reduces the energy requirements typically associated with water treatment and subsequent distribution processes. On the flip side, adopting energy-efficient protocols can tangibly curtail water consumption, particularly in heavy energy consumer sectors, like power generation [48]. Hence, understanding and capitalising on this intertwined relationship between energy and water systems can pave the way for a holistic and effective strategy for sustainable resource management.

5.4 DSM in energy systems

DSM, when viewed within the landscape of energy systems, offers an intriguing perspective on how energy is consumed. Specifically, this approach is centred around methodologies that utility services employ, aiming to inspire and encourage users to reevaluate and potentially change their energy consumption behaviours and patterns [1]. A closer examination of the DSM reveals an underlying emphasis on adjusting or decreasing demand. This presents a marked shift from more conventional energy methodologies, which mainly focus on increasing supply to meet demand [8].

Delving into the annals of history, one would find that the concept of DSM was not born in a vacuum. The seeds for this approach were sown during the energy crisis of the 1970s. This period, marked by significant tumult and uncertainty, underscored and magnified the importance of energy conservation, thereby firmly embedding it within the broader framework of energy policy and planning [9]. Fast forward to more recent times and one can observe the undeniable influence of technological advancements. From the advent of various renewable energy sources to growing global apprehensions about climate change, these developments have propelled DSM to the forefront of contemporary energy dialogues [6]. At its core, DSM's objectives in the context of energy can be delineated into a few critical goals. One of the cardinal objectives of DSM is to restrain or limit peak demand. This, in its essence, reduces the pressing need to ramp up generation capacity, thus preventing undue strain on energy infrastructures. But DSM continues beyond mitigating peak demand. There is an intricate dance between energy supply and demand, and DSM aims to choreograph this dance to perfection. By shifting energy consumption toward off-peak times, DSM fosters a harmonious alignment between the available energy and the consumed energy [43]. The overarching aspirations of DSM also include enhancing overall energy efficiency, effectively reducing total energy consumption. Furthermore, DSM sets the stage for renewable energy sources, facilitating their smooth integration into established grid systems and ensuring that they seamlessly coexist with traditional energy sources [49].

Given the depth and breadth of the DSM approach, it is common that many strategies fall under its expansive umbrella. Iqbal *et al* [49] categorises these strategies primarily into energy efficiency, DR, and load management. Let us delve a little deeper into these categorisations. Initiatives focusing on energy efficiency are meticulously crafted to reduce energy consumption without compromising the quality or quantum of services provided. Such strategies often champion using advanced, high-efficiency appliances or state-of-the-art industrial machinery. Venturing into DR initiatives, these are uniquely tailored to offer tangible incentives to consumers. The objective? To persuade them to limit or reduce their energy consumption when demand surges. This often involves dynamic pricing structures, where the price of energy spikes with its demand. Another tool in the DR arsenal is DLC, a method where utility providers can directly influence and limit the energy consumed by specific appliances, especially during demand surges [30]. While load management may echo facets of DR, it predominantly zeros in on shifting energy consumption from peak periods to off-peak windows. The focus here is less on overall reduction and more on smart redistribution [50].

Several technological tools and innovations have emerged as quintessential enablers to propel DSM into effective action. For example, the advent of smart metres allows consumers and providers to access real-time data on energy consumption. Smart grids, another innovation, show remarkable adaptability, adjusting in real-time to the ebb and flow of energy demand. Additionally, the energy landscape has seen the development of advanced energy storage mechanisms that can store surplus energy for future use. To top it all, sophisticated data analytics tools have been designed to scrutinise, understand, and even predict energy consumption patterns, ensuring that DSM strategies are proactive and effective [6].

As with any intricate system, DSM is not the sole prerogative of a single stakeholder. Instead, it functions within a web of interconnected entities, each playing a crucial role in shaping its narrative. Consumers, the end-users of energy, are at the heart of DSM as they adjust and adapt their consumption behaviours in response to various stimuli. Utility providers, the entities responsible for supplying energy, orchestrate and diligently manage various DSM programmes, ensuring they align with overarching objectives. Then there are the regulators, the watchdogs of the energy world. Their role involves meticulous oversight, ensuring that DSM policies and programmes are smoothly implemented and adhere to established norms. Finally, third-party energy service companies have carved a niche within the DSM framework. These companies offer various services and solutions custom made for DSM, ensuring the approach remains dynamic and responsive [46]. With their unique contributions, each of these stakeholders plays a role in the larger DSM tapestry, weaving together a comprehensive ecosystem that underscores the essence of sustainable energy management [50].

5.4.1 Energy efficiency initiatives and programmes

Energy efficiency initiatives and programmes have created a significant niche within the overarching framework of DSM. The influence of such programmes is felt in myriad ways, most notably in the manner in which energy consumption patterns are moulded and also the broader enhancement in the utilisation of energy resources [1]. An observer could discern the deep-seated nexus between DSM and energy efficiency. These are not just two isolated entities; rather, energy efficiency measures are intrinsically laced with the core objectives of DSM. Such a profound interconnection is evident in a variety of initiatives that focus intently on curbing the peak demand for energy, stymieing wasteful energy practises, and substantially reducing the ecological ramifications of energy consumption [8].

Delving deeper into the realm of DSM, one would find it teeming with many energy efficiency initiatives. Each initiative, distinct in its design and goals, addresses different aspects of energy consumption. For instance, energy-efficient appliances programmes prominently stand out. Central to these programmes is the ardent advocacy for appliances designed to be economical in energy consumption. This advocacy is not merely a call to action; it is not backed by tangible incentives such as rebates and discounts. Furthermore, some regions have taken this a step further by enshrining mandatory appliance efficiency standards. The ripple effects of such proactive strategies extend beyond energy savings. Indeed, the environment, too, reaps the benefits, most palpably in the form of diminished greenhouse gas emissions, which are notorious for exacerbating climate change [3].

Yet, the ambit of energy efficiency is not circumscribed to appliances alone. It is an expansive field that can zero in on buildings—the edifices that dot urban landscapes, ranging from homes to towering office complexes. Termed as building efficiency programmes, these initiatives dissect the intricate energy dynamics ensconced within residential, commercial, and public structures. At the forefront of these programmes are legislated energy-efficient building codes. These are not just pieces of legislation gathering dust; they are actively complemented by efforts to infuse buildings with energy-conserving amenities. Whether it is the heating, ventilation, or air conditioning (HVAC) systems, the insulation materials, lighting fixtures, and many other components are scrutinised and optimised to ensure minimal energy wastage. And, like their appliance counterparts, these building-focused initiatives culminate in substantial energy savings and a commendable reduction in carbon emissions, furthering the cause of environmental preservation [11].

Shifting the lens to the buzzing hubs of production and commerce, industries and businesses emerge as pivotal components in this energy efficiency mosaic. These vast sectors, known for their huge energy appetites, are addressed through the industrial and commercial efficiency programmes. At the heart of these programmes lies a thorough energy audit. These are not cursory examinations; they dive deep, identifying potential chinks in the energy armour and pinpointing areas ripe for efficiency enhancement. These audits are not conducted in isolation. They are paired with attractive financial incentives that coax these industries to embrace cutting-edge, efficient technologies. Furthermore, there is a concerted push towards educating these sectors, instilling practises that champion prudent energy consumption in them. The dividends from such focused initiatives are multiple. Apart from the evident dip in energy expenditures, there is the invaluable benefit of minimal ecological degradation [6].

In the grand tapestry of energy efficiency, financial and regulatory incentives emerge as the golden threads, weaving together disparate elements and catalysing a shift towards more energy-efficient paradigms. Whether these incentives emanate from government funds or are proffered by utility companies, they take various forms. From enticing tax reductions to generous rebates and grants, these fiscal incentives act as magnets, drawing stakeholders towards investing in energy-conserving technologies. Complementing these financial levers is the power of regulatory frameworks. Among these, the mandates dictating minimum energy performance standards stand out, as they not only set benchmarks but actively influence market dynamics, pushing it inexorably towards energy-efficient products and services [5].

The current epoch, dominated by rapid technological advancements, has seen energy efficiency programmes derive tremendous leverage from state-of-the-art innovations. Technologies like smart grids and advanced metering infrastructure are not just fancy terminologies; they have tangibly revolutionised the energy sector. Their most salient feature is the ability to meticulously monitor energy usage in real-time. Such granular insights empower both utility providers and consumers, allowing them to fine-tune and optimise their energy consumption patterns [43]. Beyond these monumental infrastructural innovations, a continuous stream of energy-conserving technologies makes their presence felt. From appliances that sip energy to LED lighting solutions, from elite HVAC systems to groundbreaking control mechanisms, each technological marvel significantly increases energy efficiency levels.

However, in pursuing energy efficiency, it is imperative to periodically pause and reflect, evaluating the outcomes of these multi-faceted endeavours. This introspection involves a meticulous assessment of the impact of energy efficiency programmes. A suite of techniques is employed for this evaluative exercise. Quantifying energy savings, for example, ranges from straightforward engineering estimates to complex analysis of metered data, supported by statistical tools. Furthermore, the economic viability of these ventures is taken with a grain of salt. Rigorous benchmarks, such as the Total Resource Cost test or the Societal Cost Test, are used to scrutinise the cost-effectiveness of these energy efficiency endeavours [5].

5.4.2 Peak load shifting strategies and ToU pricing

Energy consumption patterns are similar to the ebb and flow of tides. Just as the tides rise and fall, energy demand experiences its crests and troughs throughout the day, with particular moments that show pronounced spikes. Such moments, characterised by heightened energy demand, are aptly termed 'peak demand' [1]. This phenomenon, although seemingly straightforward, comes with its share of challenges for utility providers and consumers alike. Given the volatility of these consumption patterns, the energy sector has had to devise strategies to maintain equilibrium. Among these strategies, peak load shifting stands out as a salient solution. It operates on a simple, yet ingenious, premise—reallocating energy consumption from times marked by its acute demand to those periods when demand slackens. By doing so, it plays an instrumental role in maintaining the stability of the energy grid and ensuring that energy is utilised optimally, a feat integral for preserving resources and promoting overarching efficiency [8].

In an effort to actualise peak load shifting, innovators and policymakers alike have created a range of techniques. One such notable technique is DLC, which provides utility companies with the ability to exercise remote control over the power consumption of specific appliances, particularly during those periods marred by excessive demand [3]. But what happens when the demand subsides? That is where thermal energy storage steps in. This method focusses on harnessing and storing thermal energy when demand is at a nadir, only to release and utilise it when the demand scales its zenith. This principle of storage and timely utilisation is not confined to thermal energy alone. Battery energy storage systems adopt a similar philosophy. They enable the stowing away of electrical energy when demand is relatively modest, preserving it for those critical moments when consumption spikes occur [6].

But it is not all about control and storage. The role of consumers in this intricate dance of demand and supply cannot be understated. Recognising this, DR programmes have been meticulously designed. These programmes are not about dictating terms to consumers. Instead, they seek to collaborate. By offering incentives, they encourage consumers to rein in their energy consumption during those times when demand is soaring. This could manifest in the form of dynamic pricing mechanisms that reflect real-time energy demand or timely notifications that inform consumers of peak periods, urging them to modulate their consumption habits [15].

The advantages arising from these myriad strategies are diverse and profound. At a macro level, they potentially prevent the need for heavy investments in energy infrastructure, thus alleviating financial burdens on utility providers. In addition, they enhance the reliability of the energy grid, ensuring uninterrupted power supply even during challenging peak demand periods. Additionally, there is a considerable environmental dividend. By tempering the reliance on peaker plants, which are conventionally activated to cater to heightened demand, there is a tangible reduction in greenhouse gas emissions. These environmental gains also cascade into economic ones. Utility providers enjoy cost savings, a benefit that often trickles down to consumers in the form of reduced energy bills [43].

In tandem with these load-shifting strategies, the energy sector has witnessed the rise of a potent pricing tool, ToU pricing. As the name suggests, ToU pricing is not static. It fluctuates, modulating energy prices on the specific times of consumption. In essence, it segments the day into distinct intervals: off-peak, mid-peak, and peak, with each interval having unique pricing [37]. This dynamic pricing model comes packed with a plethora of benefits. First, it acts as a nudge, prompting consumers to rearrange their energy-intensive activities to periods designated as off-peak, ensuring both cost savings and efficient energy utilisation. Furthermore, it works seamlessly with renewable energy. For instance, solar energy can be optimally harnessed during sun-drenched hours, aligning with favourable pricing intervals. This makes renewable energy more viable and reduces the need to rely on environmentally harmful peak power plants [43]. However, like all strategies, ToU pricing is not without its challenges. Its implementation needs to be nuanced, ensuring it does not inadvertently impose financial burdens on low-income households. Moreover, its success hinges on the deployment of advanced infrastructures such as smart metres. There is also the Herculean task of spurring behavioural shifts amongst consumers, overcoming entrenched habits to adapt to new consumption rhythms [30].

Zooming out to survey the expansive energy landscape, one would notice a fascinating interplay between peak load shifting and ToU pricing. Their relationship is not one of isolated silos; it is symbiotic. For instance, ToU pricing, with its dynamic nature, can act as a potent financial impetus. It can inspire and motivate consumers to adjust their energy consumption behaviours, aligning them with the principles of peak load shift. It is a relationship where one complements the other, ensuring that the energy sector remains both sustainable and efficient [51]. As energy demands continue to evolve, the synergy between strategies such as peak load shifting and innovative tools such as ToU pricing will remain indispensable in charting a sustainable energy future.

5.4.3 DR and load curtailment techniques

In the realm of energy management, DR and load curtailment strategies emerge as two critical pillars. Both are foundational to the philosophy of DSM, which seeks to optimise energy consumption patterns in sync with the demands and capacities of the power grid [1]. The nuanced approaches offered by these strategies do not merely hover around the realm of theory, but profoundly shape how energy is consumed, distributed, and optimised in the real world. Delving deeper into DR, one uncovers a sophisticated mechanism that reacts to and manoeuvres the electric consumption behaviours of end-users. At its core, DR represents deviations from established norms of energy consumption. These deviations are not arbitrary but are typically spurred by external stimuli—sometimes the fluctuating prices of electricity, and at other times, driven by the allure of monetary incentives [3]. In this sense, DR carves out a distinctive niche for itself within the broader gamut of DSM strategies, primarily due to the unique triggers that set it into motion.

DR helps alleviate peak demand stress on the grid, reducing the risk of blackouts and brownouts during periods of high electricity consumption [27]. Therefore, it improves the stability and reliability [52]. By encouraging consumers to move their consumption to off-peak periods, the response to demand reduces wasteful energy use, leading to overall improvements in energy efficiency [53]. DR complements the variable nature of renewable energy sources such as solar and wind by aligning energy consumption with its availability, hence encouraging renewable energy integration [54–56]. Consumers who participate in DR programmes can benefit from lower electricity rates during off-peak periods, leading to savings on their energy bills [21]. The benefits and applications of the DR programme in the energy sector are encompassing to achieve efficiency and sustainability. However, widespread implementation requires three things: (i) technological readiness: advanced metering and communication infrastructure, which may be lacking in some regions [57]; (ii) consumer awareness and engagement: consumer education and awareness campaigns are necessary to effectively promote participation in DR programmes [58]; (iii) regulatory and policy barriers: complex regulations and utility policies can impede the development and implementation of DR initiatives.

The ripple effects of DR in an energy grid are profound. For starters, it bolsters the grid's resilience, ensuring that it can withstand and adapt to varying demand pressures. By encouraging shifts in consumption patterns, DR plays a pivotal role in attenuating peak demand, thus preventing the grid from being overburdened during periods of intense energy consumption [51]. But the implications are not just infrastructural. There are tangible economic advantages that cascade down to both the end-users and the utility providers. Reduced strain during peak times implies less wear and tear and a diminished need for infrastructural upgrades or repairs.

However, it would be an oversimplification to label DR as a monolithic strategy. In reality, it is a mosaic of diverse programmes, each tailored to address specific facets of energy consumption. On the one end of the spectrum, we have price-based initiatives. ToU, for example, modulates pricing based on predetermined times, urging consumers to adjust their energy usage accordingly. Critical peak pricing (CPP) sharpens this focus, dynamically increasing prices during periods of extreme grid stress. Then there is real-time pricing (RTP), which provides users with pricing that mirrors real-time grid conditions, thus offering them the flexibility and information to adapt their consumption patterns in sync with grid health [6]. On the other hand, incentive-based DR strategies lure consumers with potential financial rewards. DLC represents a more direct intervention, allowing utilities to remotely control certain appliances. The interruptible load, meanwhile, gives consumers the option to reduce or halt their consumption in return for compensations. CAP programmes function by securing commitments from users to reduce their load during certain conditions, effectively creating a buffer against unforeseen spikes in demand.

The collective might of these programmes plays a symphonic role in ensuring that the energy supply aligns gracefully with demand. The environmental dividends are also palpable, as optimised usage reduces greenhouse gas emissions and lessens the dependence on peaker plants, which typically kick in during high-demand phases. However, DR's journey is not devoid of challenges. To unleash its full potential, a robust technological foundation, epitomised by advanced metering systems, is a prerequisite. Furthermore, the success of DR is inextricably linked to consumer awareness, engagement, and participation. The strategy also must navigate the intricate terrains of communication and control technologies, ensuring seamless interactions between the grid, utility providers, and consumers [51].

Transitioning from DR, the world of load-curtailment techniques offers another rich tapestry of strategies. The underlying philosophy here revolves around reducing or adjusting energy loads, especially during times when the grid is stretched to its limits or faces unexpected challenges [5]. Taking a leaf out of DR's book, DLC in this context empowers utilities to remotely influence specific appliances, a tactic that can be especially handy during crunch periods. Another technique, scheduled load curtailment, adds an element of predictability. By setting predetermined time slots and inviting consumers to reduce their consumption during these periods, it ensures a planned reduction in the grid load. However, not all load curtailment hinges on schedules or remote controls. Voluntary load curtailment bestows upon consumers the autonomy to modulate their consumption, usually spurred by financial incentives or price cues.

A somewhat divergent strategy within load curtailment is distributed generation. Instead of focusing on reducing consumption, this approach promotes the generation of energy at localised points. Think of small-scale solar installations on rooftops or small wind turbines that dot a landscape. Such localised energy sources can feed into the grid, alleviating its burden, especially during times of duress [13]. Much like DR, the load curtailment techniques also promise a bouquet of benefits. By regularly fine-tuning consumption patterns, they fortify grid stability. They also play a pivotal role in reducing the costs associated with peak power generation, ensuring that the grid does not frequently resort to expensive peaker plants. Furthermore, these techniques pave the way for a smoother integration of renewable energy sources into the grid. However, the path is not always smooth. Consistent engagement of consumers, ensuring they collaborate in these initiatives, remains a significant challenge. The need for sophisticated technology is ever present, and there is always the looming shadow of potential inconveniences that consumers might experience in the process.

In conclusion, the intertwined worlds of DR and load curtailment techniques, nestled within the broader framework of DSM, are testament to the dynamic, innovative, and nuanced ways in which energy consumption is being reimagined and restructured in the contemporary era. Through their diverse strategies and programmes, they seek to craft an energy landscape that is not only efficient and resilient but is also harmoniously in sync with the environment and the evolving needs of consumers.

5.4.4 Integrating renewable energy and DSM

The world's energy sector is currently undergoing a significant metamorphosis, pivoting towards more sustainable paradigms. This transition is becoming increasingly evident, with the prominence of DSM and renewable energy becoming more pronounced [12]. The global narrative on energy is being reshaped, with DSM emerging as a pivotal mechanism that ensures that the energy grid operates efficiently and demonstrates resilience against potential challenges. Concurrently, the rise of renewable energy, underlined by power sources like wind, solar, and hydro, has highlighted the importance of integrating these renewable sources with the more traditional energy systems [10].

Diving into the world of renewable energy reveals a plethora of sources, each unique but collectively contributing to a more sustainable energy landscape. Integration into the primary energy grid systems offers a host of benefits. One of the most salient advantages is the significant reduction in greenhouse gas emissions. This not only represents an environmental boon, but also positions the energy sector as a key player in global efforts to combat climate change. In addition, the integration of renewable sources into the grid underscores the concept of energy security. As nations become less dependent on fossil fuels, the geopolitical landscape surrounding energy may see considerable changes. However, while the potential benefits of renewable integration are significant, the challenges are equally pronounced. Solar and wind, two of the prominent renewable energy sources, are inherently contingent on natural factors, such as the availability of sunlight and wind speeds. Such variability is problematic, resulting in energy output fluctuations that the grid must constantly accommodate.

While renewable energy is becoming an integral part of the energy matrix, DSM operates in the backdrop, ensuring that the grid remains stable and efficient. DSM represents a strategic approach to energy management, focussing on system reliability, managing peak demand, and generating economic returns [10]. The essence of DSM lies in its multi-faceted strategies, which include DR, load curtailment, and several others. These strategies act as regulators, fine-tuning energy demand to ensure it aligns with supply, thereby ensuring that the grid remains balanced and operational.

When one examines the relationship between renewable energy and DSM, the synergies become abundantly clear. The techniques and strategies within DSM can effectively counterbalance the inherent variability of renewable energy sources, leading to a more synchronised and harmonious energy ecosystem [10]. This harmonisation is especially evident in the context of DR. DR strategies can be tailored to ensure that energy consumption patterns are adjusted to reflect the availability of renewable energy, thereby ensuring that energy demand spikes do not occur during periods of low renewable output. An additional layer of complexity and capability is added with the advent of energy storage solutions, especially battery technologies. Batteries can store excess energy generated during peak renewable output periods, and then release this stored energy during periods of high demand or low renewable output. This functionality essentially bridges the objectives of DSM with the capabilities of renewable energy systems [37].

It is undeniable that technological advancements are the primary drivers behind the seamless integration of renewable energy and DSM. Advanced metering infrastructure (AMI) provides invaluable insights into energy consumption at a granular level. This information allows for more adaptive and flexible demand management [59]. Simultaneously, energy management systems play a vital orchestration role, ensuring that the interaction between renewable energy sources and DSM is seamless and efficient [60]. With the rapid proliferation of the Internet of Things (IoT), real-time energy monitoring and management become possible. The array of connected devices, sensors, and systems provide a continuous stream of data, which, when combined with artificial intelligence and machine learning, can be used for predictive analysis. Such analysis allows for the anticipation of energy consumption patterns and the optimisation of both supply and demand to ensure that the grid remains stable [4].

However, it would be an oversight to assume that the integration of renewable energy and DSM is without its challenges. From a technical point of view, ensuring grid stability in the face of variable renewable output, accurately forecasting energy loads, and upgrading existing infrastructure to accommodate new technologies are significant tasks [61]. There are also considerable economic challenges to consider. These encompass the costs associated with transitioning to renewable energy, the implications of energy storage, and the capital investments required for DSM systems [4]. In addition, the regulatory frameworks that govern the energy sectors around the globe often struggle to keep up with rapid technological advancements, resulting in potential impediments to integration. Lastly, and perhaps most unpredictably, the behaviour and adaptability of consumers play a decisive role in the success of integrating renewable energy with DSM. Whether it is adjusting to new pricing models, adopting new technologies, or simply being receptive to changing energy consumption patterns, the human element remains both a challenge and an opportunity in this evolving energy landscape [30].

5.4.5 Case studies of successful implementation of DSM in energy systems

The integration of case studies from various global endeavours provides a rich tapestry of insights into the art and science of implementing DSM strategies. Examining these efforts from different geographies and regulatory environments illuminates the collective strides made toward the sustainable energy transition and offers a roadmap for best practises, potential pitfalls, and innovative solutions.

Take, for instance, Germany's Energiewende, an endeavour that stands out as a beacon of ambition in the realm of sustainable energy transitions. The German nation charted an audacious course toward establishing a predominantly renewable energy ecosystem. The target was set: achieve a staggering 80% renewable energy by 2050. This was not a mere aspiration; Germany began to put the pieces in place by leaning heavily into investments in wind, solar, and biomass energy. To further underscore its commitment, in the aftermath of the Fukushima disaster, the nation took the bold step of shutting down their nuclear power plants. The results of these decisions were palpable. By the time the calendar flipped to 2020, Germany had approximately 42% of its electricity generated from renewable sources [7]. Beyond the sheer numbers, what this transition effectively showcased was the transformative power of policy incentives. One of the standout policy manoeuvres was the introduction of feed-in tariffs, which played a crucial role in accelerating the adoption of renewable energy sources.

Drawing parallels from another part of the world, Hawaii embarked on its renewable energy journey, committing itself to a future powered 100% by renewable energy by 2045. This was not a journey they embarked on in isolation. Hawaii actively promoted a culture of promoting investments in wind and solar energy. A salient feature of their strategy was the emphasis on adopting solar adoption. This emphasis was not in vain; Hawaii observed a meteoric rise in the adoption of household solar energy solutions [2]. An underlying theme of Hawaii's energy transition narrative was the role of community. The state's experience highlighted the profound impact of community engagement in renewable energy projects, emphasising the collective spirit and shared responsibility in sustainable energy adoption [4].

The United States presented another captivating chapter in the annals of DSM. PJM Interconnection, which functions as a regional transmission organisation, envisioned and subsequently introduced DR programmes. The primary goal was straightforward: improve the reliability of the grid. The dividends from this initiative were almost immediate. During periods of peak demand, particularly when faced with the vagaries of extreme weather conditions, PJM Interconnection documented savings that were in the millions. However, this accomplishment was not solely due to policy implementation. It shone a spotlight on the indispensibility of robust data infrastructure and the pivotal role of consumer participation in ensuring the success of such programmes [1].

Australia's energy landscape was grappling with the rising tides of energy demand, a situation exacerbated during their sweltering summer peaks. The solution was born in collaboration. Energy providers and consumers found common ground, introducing RTP as a tool to curtail consumption. The outcomes spoke volumes, as there was a marked reduction in load, especially during those critical peak periods [7]. Australia's experience served as a testament to the effectiveness of flexible energy pricing and how it can reshape consumption behaviours [11].

Further afield, California championed the concept of ToU rates. The primary objective was to synchronise consumption habits with the availability of renewable energy. By supporting variable pricing, dependent on energy demand, and coupling it with comprehensive public awareness campaigns, California witnessed a paradigm shift in consumption patterns. This shift was instrumental in furthering the integration of solar energy [59]. The state's journey unravelled two key insights: first, the indispensable role of consumer education in shaping energy consumption behaviours, and second, the need for synergistic collaboration between utility providers and regulators [62].

Ontario, Canada, presented a unique challenge and solution dichotomy. With the objective of reducing peak loads, especially during chilly winter evenings and scorching summer afternoons, Ontario rolled out off-peak incentives. Through a strategy that championed differential pricing and encouraged energy-intensive activities during off-peak times, the province witnessed a discernible shift in energy consumption patterns. This strategy not only alleviated pressure on the grid but also accentuated the importance of effective marketing in shaping and implementing pricing strategies [62]. This initiative echoed the sentiments of [59] regarding the effectiveness of marketing in energy transition strategies.

Japan, on the other hand, adopted a different approach. The nation's Top Runner Programme was laser-focused on elevating the energy efficiency of appliances. By setting benchmarks based on the best-performing products in the market and revisiting these benchmarks periodically, Japan realised significant advancements in appliance energy efficiency [63]. This endeavour showcased that dynamic targets, which evolve based on market performance, could potentially offer greater efficacy compared to static benchmarks [64].

Wrapping up this global tour, the European Union embarked on an ambitious mission, striving to improve energy efficiency by 32.5% by 2030. The strategy was comprehensive, encompassing legally binding measures imposed on member states, promotion of building retrofitting, the introduction of efficient appliances, and the promotion of sustainable industrial practises. The results were evident, with the union making significant strides towards their 2030 objective. The European Union's methodology underscored two core principles: regional collaboration can amplify impact, and constant monitoring and enforcement mechanisms are critical to ensure compliance and progress [50].

In summation, the global tapestry of DSM strategies and renewable energy initiatives offers a mosaic of insights, challenges, and solutions. Each geography, with its unique challenges and resources, has created a path that others can learn from. Whether it is the policy-driven approach of Germany, the community-centric strategy of Hawaii, or the data-driven methodologies of PJM Interconnection, each case presents a chapter in the larger narrative of global sustainable energy transition.

5.5 DSM in water systems

In the context of water systems, DSM is characterised as strategies and actions designed to control, adjust, or reshape consumer demand for water [64]. Instead of the traditional approach which primarily focusses on increasing water supply—via methods such as establishing new reservoirs, desalination plants, or inter-basin transfers—DSM prioritises efficient water utilisation, conservation, and modification of water-use behaviours at the end-user level [50].

The increasing water scarcity underscores the global imperative for DSM in water management challenges many regions face. This is due to factors such as excessive water extraction, shifts in climatic patterns, and burgeoning populations [16]. Conventional methods to improve water supply, like a dam or reservoir construction, often involve substantial financial expenditures and can lead to ecological and societal complications. Coupled with the unreliability of rainfall patterns, predominantly attributed to climate change, exclusive reliance on these traditional strategies becomes increasingly precarious [3].

The rise in urbanisation further compounds these challenges. As cities burgeon, so does the water demand, yet water sources remain invariably limited [16]. Here, DSM comes into play, offering a more sustainable approach to managing water demand. By implementing public awareness initiatives, revised pricing structures, and introducing water-efficient technologies, DSM enables communities to optimise every litre of water consumed, minimising the dependency on capital-intensive and potentially detrimental infrastructural developments [6].

In conclusion, DSM serves as an essential framework for addressing contemporary water dilemmas, accentuating the need for prudent water resource management [9]. DSM fosters a conservation-centric ethos by redirecting attention from solely enhancing supply to effectively governing demand, safeguarding this critical resource for subsequent generations.

5.5.1 Demand management strategies in the water industry

Demand management within the water industry is fundamentally about strategies and actions to control and manage water consumption [64]. Given the backdrop of escalating water scarcity, booming global populations, and the omnipresent threats of climate change, there is an urgent imperative to ensure water's sustainable use. A prominent strategy employed is the use of nuanced pricing mechanisms. Tiered pricing, for instance, escalates the costs as consumption increases, encouraging users to be more judicious with their water use [1]. However, the overarching challenge remains to ensure fairness across different income groups. On the other hand, seasonal pricing modifies prices based on the water availability of a particular season, providing a disincentive on wastewater during dryer periods. The unpredictability of climate, however, makes it challenging to forecast seasonal changes accurately [3]. Furthermore, differential pricing establishes varied price points for domestic, industrial, and agricultural consumption. Although this approach can effectively target the most significant consumption sectors, the difficulty lies in setting differentials that do not hinder economic activity [40].

Public awareness and education are the bedrock of demand management. By weaving water conservation themes into school curricula, there is an opportunity to instill conservation values early on [1]. Town hall meetings and workshops foster direct community participation, often leading to improved understanding and adherence to conservation practises. When employed effectively, media campaigns can reach large audiences and cater to specific demographic targets, further amplifying the conservation message [6]. From a technological point, adopting low-flow appliances, such as toilets, reduces water use. However, challenges arise concerning the initial investment and market availability [40]. Smart metering systems, which relay real-time water consumption data, enhance consumer awareness and are instrumental in early leak detection. Yet, these systems require significant infrastructure investment, and there is an overarching concern about data privacy. In the agricultural sector, the move towards more water-efficient irrigation methods, such as drip systems, has the potential for vast water savings. However, the initial setup costs and getting farmers on board remain substantial hurdles.

On the regulatory front, mandates that require water-efficient appliances in new constructions offer a path to standardising efficiency across buildings, although the challenges of monitoring and compliance persist [5]. Enforcing specific water use can rapidly curb demand, especially during critical periods, but these often meet with public resistance and are difficult to enforce. Offering subsidies for water-efficient appliances can expedite their adoption, yet the challenge lies in securing funds and ensuring that these incentives drive genuine uptake [6]. Lastly, exploring alternative water sources like rainwater harvesting and greywater recycling has gained traction. Capturing rainwater reduces dependence on primary water sources, although concerns about initial costs and water quality persist [64]. Greywater recycling, which involves reusing lightly contaminated water, can drastically cut freshwater demand for non-drinking purposes. However, this raises issues around health implications and the requisite infrastructure [3].

5.5.2 Water conservation programmes and water-efficient technologies

Sustainability in water usage requires a harmonious blend of effective conservation programmes and the integration of water-efficient technologies [1]. DSM emphasises public awareness campaigns employing media outlets, including TV, radio, and digital platforms, to accentuate the seriousness of water scarcity and the conservation imperative. Such campaigns have led to a notable reduction in water consumption after successful campaign runs [2].

In addition, DSM places significance on both school and community participation. Incorporating water conservation topics into curriculums, coupled with community workshops and challenges, promotes conservation habits. Such initiatives have been pivotal in catalysing long-term behavioural change and strengthening community cohesion around water-saving practises [35]. Another cornerstone of DSM is the implementation of tiered pricing mechanisms and incentives. By instituting higher costs for escalated consumption, offering rebates for reduced usage, and incentivising the adoption of water-saving technologies, there is observed reduced wastage and a surge in the uptake of efficient technologies. Regulations, too, play a crucial role; by mandating the inclusion of water-efficient fixtures in new constructions and setting water use restrictions during shortages, there is an ensured, standardised deployment of efficient technologies, leading to instantaneous demand mitigation during crises [5].

In the realm of water-efficient technologies as part of DSM, low-flow appliances stand out as a quintessential example. These devices, which include low-flow toilets and showerheads, are engineered to utilise minimal water without compromising efficacy. Their adoption translates into a direct contraction in household water consumption [59]. The introduction of smart water metres, which relay real-time water usage data, has improved consumer awareness, making them instrumental in early leak detection. Technologies such as drip irrigation and soaker hoses have revolutionised the agricultural sector, substantially reducing water wastage without detriment to crop yields. Furthermore, greywater recycling systems, which reuse water from domestic utilities like sinks and showers for non-potable applications, alleviate the strain on freshwater sources, rendering them particularly advantageous in arid regions [3]. Similarly, rainwater harvesting systems that capture and store rain for subsequent utilisation ease the burden on the municipal water supply and offer a cushion during rain-deficient phases [64].

5.5.3 Sustainable agricultural practises and water use

A vital component of the DSM framework is the integration of sustainable agricultural practices by optimising irrigation practises [6]. Agriculture accounts for a substantial portion of global freshwater usage; about 70% of global freshwater withdrawals are directed toward agriculture. This usage poses challenges such as over-irrigation, water wastage, and rapid depletion of aquifers [59]. Sustainable agricultural practises come into play to mitigate these problem. Smart irrigation systems equipped with sensors and real-time data monitoring allow farmers to adjust irrigation schedules based on weather conditions, soil moisture levels, and water availability [43]. By aligning irrigation with periods of lower water demand, farmers can conserve water resources and reduce energy consumption associated with pumping water [65]. Crop rotation, which involves the systematic rotation of crops on specific land plots, not only bolsters soil health and reduces pest infestations, but it also optimises water usage [34]. Agroforestry, which integrates trees within crop and animal farming, is another pivotal practise; it bolsters water retention, minimises evaporation, and facilitates increased groundwater recharge [13]. Additionally, planting cover crops specifically to protect and enrich the soil—rather than for harvest—kerbs soil erosion, amplifies soil moisture, and diminishes the need for fertilisers [5]. Organic farming, which eschews synthetic pesticides and fertilisers, paves the way for healthier, water-retentive soil while concurrently slashing water pollution. In addition, conservation tillage moderates the frequency and intensity of tilling, preserves soil structure, and enhances water retention while reducing runoff [13].

Complementing these sustainable practises are DSM's water-saving technologies for agriculture. Drip irrigation systems, which deliver water right to the base of a plant, promise significant water savings, reduced evaporation, and optimised nutrient use [66]. Soil moisture sensors, which gauge soil moisture levels, ensure that plants receive just the right amount of water, averting over-irrigation [3]. Furthermore, weather-based irrigation controllers, which modify irrigation schedules in alignment with weather conditions, facilitate precision irrigation, culminating in considerable water savings [5]. The DSM is seminal in fostering sustainable agriculture by championing policies and incentives. Governments and institutions progressively implement initiatives to incentivise these sustainable practices. It is also essential to educate farmers about the multi-faceted benefits of sustainable practises. Furthermore, sustainable farming is ever evolving, underscoring the indispensability of continued investments in research and development to uncover and refine sustainable farming techniques and technologies [5].

5.5.4 Smart water metres and consumer engagement

Water conservation has become paramount due to many global challenges, such as climate change, population growth, and diminishing water resources [6]. Technological advancements play a crucial role in the arsenal of strategies to address these challenges, and smart water metres are emerging as a crucial tool. These metres bridge the gap between traditional water management and modern consumer engagement, catalysing efficient water management [31]. Historically, water management relied heavily on older metering systems with several limitations, including inaccurate readings and manual data collection [67]. But with the onset of the digital age, the water industry saw a transformation. Smart water metres, with capabilities like real-time monitoring, wireless data transmission, and remote control functions, entered the scene [60].

For utilities, these metres offer a plethora of advantages. They enable accurate and timely billing, facilitate early leak detection within the system, and support better demand prediction and management. Conversely, consumers enjoy detailed insights into their water consumption patterns, allowing them to pinpoint household leaks early on and potentially accrue savings through more efficient water use [68]. Furthermore, smart metres amplify consumer engagement in multiple ways. They provide real-time feedback, allowing users to quickly understand and adjust their water usage patterns [64]. Utility can generate personalised consumption reports, helping consumers establish and meet their water conservation goals. Additionally, with incentive programmes, utilities can offer a carrot of rewards or rebates for households exhibiting efficient water usage patterns. Another intriguing facet are community comparisons. By juxtaposing one's consumption with neighbours or community averages, a sense of friendly competition can be sparked, fuelling conservation endeavours.

In the broader scope of DSM, smart metres hold an indispensable role. They allow for astute demand forecasting by harnessing real-time data, enabling effective water supply management [31]. They also play a role in peak demand management via tiered pricing models or through timely alerts nudging consumers to use water during off-peak hours [47]. Lastly, based on the granular consumption data these metres provide, utilities can proffer targeted tips and insights, further educating consumers on best practises in water use [1].

5.5.5 Water pricing policies and water use efficiency

Water, often dubbed the 'elixir of life,' is essential for human survival. However, as supplies deplete in face of the rising demands, the importance of water pricing policies in directing consumer behaviour towards efficient water use has become evident. Recognising water's increasing scarcity, there is a pressing need to attribute an economic value to it. Here, the price mechanism is an influential tool that can shape consumption patterns [47]. Among the diverse water pricing policies, flat-rate pricing offers users a fixed charge regardless of consumption, simplifying billing but inadvertently encouraging wastage. On the contrary, tiered or block pricing increases the unit price as consumption increases, promoting water savings, but potentially being perceived as punitive for larger households. Seasonal pricing, which sees rates vary by season, reflects the real cost of water supply, but its unpredictability can be challenging for households. Volumetric pricing, where charges are strictly based on consumption, is equitable and promotes conservation, but requires precise metering systems.

Pricing is not merely a consumption regulator; it communicates the value of water to consumers. A fitting price can powerfully nudge consumers towards judicious water use [68]. However, an overarching approach can prove problematic. Ensuring water accessibility and affordability, especially for vulnerable groups, is paramount. Additionally, revenues from water pricing could be reinvested into maintaining and upgrading water infrastructure, guaranteeing sustainability [47]. Elevated pricing, as seen in tiered or seasonal models, can reduce excessive consumption. Financial rewards or rebates could be introduced to acknowledge those who invest in water-efficient technologies. Furthermore, sustainability promotion, reduced rates, or incentives might be provided to those adopting water recycling systems, underlining the dual economic and ecological benefits of water reuse [64].

5.5.6 Case studies of effective DSM in water systems

In water management, various cities have adopted innovative DSM strategies for efficient water use and conservation [1]. In the arid environment of Tucson, Arizona, limited freshwater sources presented a significant challenge [6]. Addressing this, Tucson's administration introduced tiered water pricing in which costs rose with increased consumption. Along with incentives such as rebates for rainwater harvesting and high-efficiency toilets, this initiative caused a considerable decline in average household water use and instilled a conservation ethic within the community [6].

Facing a dire water crisis precipitated by extended droughts, Cape Town, South Africa, confronted the looming 'Day Zero' threat [8]. In this dire situation, the city enforced stringent water restrictions and spearheaded broad public campaigns advocating reduced daily water consumption. Collaborative efforts with local businesses to enhance water efficiency culminated in the collective effort to stave off 'Day Zero' and set new benchmarks for water consumption [8]. Constrained by its scarce natural water resources and heavy reliance on imported water, Singapore formulated a comprehensive DSM strategy [10]. This involved promoting water-efficient appliances through dedicated labelling programmes and the pioneering of NEWater, a groundbreaking treated wastewater solution for human consumption. With robust public outreach campaigns, these measures significantly curbed domestic water consumption and led to NEWater fulfilling over 40% of Singapore's water demand [10].

Recurrent droughts and arid conditions in Israel required innovative water management solutions. The country widely promoted drip irrigation, championed wastewater recycling for agricultural purposes, and invested heavily in research of drought resistant crops. This integrated approach resulted in a 50% reduction in water consumption in agriculture, firmly placing Israel at the forefront of agricultural water efficiency on a global scale. Lastly, in Barcelona, Spain, the escalating demands for water from an expanding populace prompted a tech-centric response [5]. The city started a widespread deployment of smart water metres, supported by a mobile application that allowed residents to oversee their consumption. Innovative features, such as real-time leak detection and immediate consumer notifications, underpinned this initiative, culminating in a 15% reduction in city-wide water use [5]. These diverse global examples highlight the transformative potential of well-orchestrated DSM strategies in addressing pressing water challenges [1]. Such success stories offer valuable insights for regions seeking sustainable water management solutions.

In urban water supply systems, DR can be applied to manage water storage and distribution. Smart water metres and real-time monitoring technologies enable utilities to monitor water consumption patterns and respond to fluctuations in demand [69]. Utilities can implement ToU pricing or CPP to incentivise consumers to adjust their water usage during peak hours, reducing pressure on the water distribution network. DR can be applied to water recycling and reuse systems. Industries and municipalities with water recycling facilities can adjust their water use based on the availability of recycled water [70]. During periods of high water demand or limited recycled water supply, DR measures can be implemented to prioritise water use for essential purposes and reduce nonessential water consumption. In regions experiencing droughts or water scarcity, DR may play a crucial role in water management [71]. Through educational campaigns, financial incentives, or voluntary agreements, consumers and industries can be encouraged to reduce water consumption during drought periods. By adopting water saving practices and adjusting water usage patterns, communities can collectively conserve water resources and ensure equitable water distribution [72]. DR can also be applied in industrial water management to optimise water use in manufacturing processes. Industries can adjust their water consumption patterns based on availability of water, cost, and environmental considerations. Implementing DR in industrial water management helps improve water efficiency, reduces operational costs, and improves sustainability.

In general, DR measures in water systems promote water conservation and efficiency, helping to ensure sustainable use of limited water resources. Reducing water consumption during peak hours can also save energy, as less energy is required to pump and treat water [73]. Additionally, DR in water systems enhances the resilience of water supply and distribution networks by optimising water usage and reducing strain during peak demand periods. Furthermore, by managing water use more effectively, DR can contribute to maintaining water quality and reducing the risk of contamination.

5.6 The energy–water nexus: interdependencies and challenges

5.6.1 Understanding the complex relationship between energy and water

The energy–water nexus refers to the intricate relationship between water usage in energy production and energy consumption in water processing [3]. Recognising this intertwined connection is pivotal, especially when addressing sustainable practises against evolving global challenges [1]. At the core of the energy–water nexus, the role of water in energy production manifests itself in various ways. For instance, hydropower uses water flows to generate electricity [8]. Water plays an essential role in nuclear and thermal power plants as a coolant [43]. The production of biofuels, particularly from water-intensive crops such as corn or sugarcane, further exemplifies this interdependency, as these crops require large amounts of water to produce ethanol [10]. Additionally, hydraulic fracturing, or fracking, utilises significant water quantities to extract natural gas.

Conversely, several water processes are energy intensive. The purification of water and the treatment of wastewater demand substantial energy [35]. Pumping systems, vital for the transportation and distribution of water, predominantly run on electricity [38]. Furthermore, the desalination process, which converts seawater into drinkable water, is recognised for its high energy requirements.

5.6.2 Water use in energy production: power generation, fossil fuels, and hydropower

Water is indispensable in various energy production processes, making its management crucial for sustainable energy production. In power plants, water is used primarily as a coolant, with differences existing between once-through and recirculating cooling systems [46]. On average, power plants are estimated to consume vast amounts of water, reflecting the importance of this resource in energy production. In geothermal energy, water, often from underground reservoirs, is a primary component in harnessing heat from the earth [59]. When it comes to fossil fuels, the role is equally vital. For instance, in the oil industry, water is pivotal in secondary and tertiary recovery techniques that help extract hard-to-reach oil [30]. In natural gas, hydraulic fracturing or fracking requires substantial water to fracture rocks and release gas. A single fracking operation can consume significant volumes of water. Meanwhile, coal mining extensively employs water in both the extraction of coal and its subsequent cleaning processes [6].

Hydropower, a form of energy that converts flowing water into electricity, shows another dimension of the water-energy nexus [5]. In such systems, reservoirs are essential to storing water and regulating flow. However, large reservoirs can lead to significant water losses due to evaporation, especially in warmer regions, affecting the efficiency of hydropower systems [35].

5.6.3 Energy use in water systems: water extraction, treatment, and distribution

Energy is a critical component in the process of delivering water from its source to the end consumer [1]. When groundwater is extracted, energy is primarily consumed by pumping systems designed to draw water from aquifers. The energy demand invariably increases with the depth of the water source, making deep aquifers more energy-intensive [10]. For surface water sources such as lakes and rivers, pumping stations, often located at strategic points, use significant amounts of energy to extract and move water [38]. Treatment processes to make water safe for consumption also have considerable energy implications. In potable water treatment facilities, physical processes such as sedimentation and filtration are energy consumers, with sedimentation tanks and filtration systems being particularly energy-intensive [43]. Additionally, chemical disinfection methods, including chlorination and ozonation, demand substantial energy inputs [46]. Advanced treatments, such as reverse osmosis, can further increase the energy footprint of water treatment operations [34].

Similarly, wastewater treatment processes vary in energy use. Basic sedimentation processes in primary treatment use relatively less energy compared to biological processes in secondary treatment [10]. Tertiary treatment, which often involves advanced methods for nutrient removal and further disinfection, typically has the highest energy requirements of the three stages. The distribution of water to consumers after treatment is another phase of energy-intensiveness. Pumping systems are used to push water through networks, and energy usage often depends on factors such as the distance covered and elevation changes [6]. Modern water distribution networks also incorporate monitoring and control systems, including sensors, control mechanisms, and telemetry, all of which need energy to operate [59]. Lastly, delivering water to consumers is not merely about transportation but ensuring that the water reaches homes and businesses with appropriate pressure, a task that also demands energy, especially in maintaining pressure in the system [5].

5.6.4 Impacts of climate change on the energy–water nexus

The current global climate change trends and projections are alarming, with increasing temperatures, sea level rise, and shifts in precipitation patterns being observed around the world. This change in climate has profound implications for the nexus between energy and water. Water scarcity, intensified by climate change, can pose significant challenges to energy production. For thermal power plants, which rely heavily on water for cooling, reduced water availability can hinder operational efficiency and lead to frequent shutdowns [34]. Similarly, hydroelectric generation, which depends on consistent river flows, can be adversely affected by changing precipitation patterns. Altered river flows can decrease the amount of water available for power generation, thereby reducing the reliability and efficiency of these plants [35]. Additionally, bioenergy crops, like corn and sugarcane, rely on regular precipitation patterns for optimal growth. Shifts in these patterns, as predicted under various climate change scenarios, can reduce crop yields, making bioenergy production less viable [1].

As freshwater sources become scarcer due to climate change, energy-intensive solutions could gain prominence. Desalination, a process that turns seawater into potable water, is a prime example. Although effective, it is an energy-guzzling process [3]. Similarly, treating and recycling wastewater for reuse also requires significant energy inputs, especially when ensuring that water meets potable standards. Moreover, climate change is not just about decreasing water availability. Increasing water temperatures also pose a challenge. In thermal power plants, warmer water reduces the efficiency of cooling systems, making the power generation process less efficient. On the agricultural front, warm water temperatures can affect the growth rates of bioenergy crops, potentially leading to reduced energy yields [38].

The implications of sea level rise, another consequence of climate change, are particularly concerning for coastal energy infrastructure. Power plants located along coastlines are at an elevated risk due to rising sea levels, which can lead to infrastructure damage and reduced operational capabilities. Similarly, critical oil and gas infrastructure, including offshore drilling platforms and coastal refineries, is susceptible to damage from heightened storm surges and rising sea levels [34].

5.6.5 Policy and management challenges in addressing the nexus

Addressing the energy–water nexus has gained urgency in the backdrop of escalating global challenges such as population growth, urbanisation, and climate change [1]. The water and energy sectors often operate under different regulatory bodies, resulting in fragmented policy making. There are instances where energy policies inadvertently harm water resources and vice versa, exemplifying conflicting mandates [6]. One of the significant hurdles is the need for more comprehensive data. Often, data pertinent to the nexus are scattered across various departments, creating silos that hamper effective decision-making [5]. Additionally, more consistent metrics need to be provided, particularly due to disparate units of measurement and an absence of standardised assessment metrics [35].

Regarding infrastructure, outdated systems pose considerable challenges, ill-suited for current or upcoming demands [30]. Alarmingly, there needs to be more emphasis on investing in infrastructure tailored to manage energy and water efficiently [74]. On the economic front, issues arise from subsidised pricing, which can promote wasteful consumption and misallocation of resources [4]. Financial constraints further impede the rollout of policies and technologies that consider the nexus holistically [71].

Technologically, there is a noticeable hesitance in embracing solutions that harmonise energy and water use, resulting in a slow adoption rate of integrated technologies [10]. Concurrently, there is a pressing need for research and development to devise innovative solutions tailored for this nexus. From a societal perspective, there is a general lack of understanding about the importance of the energy–water nexus, leading to unsustainable behaviours. There is also a marked resistance in modifying societal habits, a challenge when advocating for resource conservation [1].

Furthermore, policy decisions might need to pay more attention to broader ecological impacts, especially when navigating the delicate balance between energy and water demands [2]. Lastly, the shifting patterns due to climate change add another layer of complexity, further complicating policy and management efforts in the energy–water nexus realm.

5.7 Integrated approaches to energy–water nexus management

5.7.1 Coordinated strategies for sustainable resource management

Understanding the energy–water nexus involves recognising the intricate connection between energy and water systems [1]. For long-term sustainability, it is pivotal to address these resources in tandem, emphasising the importance of a comprehensive approach. The entwined nature of water and energy magnifies the complexity of conventional resource management strategies [6]. With increasing pressures from factors such as population growth, urbanisation, industrialisation, and climate change, the call for coordinated strategies becomes even more pronounced [5].

An essential aspect of this integration is cross-sectoral collaboration. It is crucial to dismantle departmental barriers and champion collective planning initiatives. Moreover, to navigate the nuances of this nexus, policies should be unified, addressing both water and energy issues concurrently, rather than through fragmented regulations [30]. On the technological front, intelligent systems, including IoT and artificial intelligence (AI), play pivotal roles in refining operations, ensuring optimal utilisation of both water and energy [10]. Furthermore, retrofitting the current infrastructure can lead to enhanced efficiency, minimising wastage across both domains [38].

One of the underpinnings for effective nexus management lies in coordinated data management. Central repositories, housing data crucial to both the energy and water sectors, can streamline decision-making processes [8]. Data-sharing agreements further this cause, urging stakeholders to disseminate information for the collective benefit [4]. Engaging in partnerships that span governments, private sectors, and NGOs can fast track sustainable solutions for the energy–water nexus. With a unified vision, these collaborations can steer the nexus towards a more sustainable trajectory.

Community participation is indispensable. Through public awareness campaigns, the masses can be informed about the nexus and its bearing on their daily lives [2]. Incentives such as rebates or subsidies could act as catalysts, fostering energy and water conservation practises among consumers [4]. From an ecological perspective, adopting strategies that resonate with the broader ecological balance is fundamental. This ensures that energy and water management endeavours do not inadvertently harm the environment [1]. Furthermore, with the ever-looming threat of climate change, embracing climate-resilient methodologies becomes imperative [5]. Lastly, regular assessments bolster the effectiveness of integrated strategies, underscoring the need for timely evaluations [4]. Feedback mechanisms can further refine these strategies, fostering adaptability based on real-time insights [1].

5.7.2 Synergies and trade-offs in DSM in energy and water sectors

The intertwining nature of energy and water sectors presents opportunities for synergies and also brings forth certain trade-offs, especially in the realm of DSM. Effective management of these synergies and trade-offs can optimise the balance between energy and water conservation, creating sustainable futures for both sectors. One of the apparent synergies lies in shared technologies. Technologies like smart metres and IoT devices can seamlessly monitor and manage resource consumption in both the energy and water sectors, providing real-time feedback and promoting efficiency [30]. In addition, unified policy frameworks could streamline conservation efforts in both sectors. Policies designed with an overarching view can yield multiplier effects in efficiency and conservation [45]. Integrated data systems are also vital. By sharing insights from both energy and water consumption patterns, these systems can support efficient resource management and decision-making processes. In addition, educating consumers about the intricacies of the energy–water nexus can engender a meaningful change in behaviour. When individuals understand the interconnected nature of these resources, they are likely to adopt practises that conserve both, yielding dual benefits [1].

On the flip side, resource allocation presents dilemmas. For instance, conserving water in hydropower dams could necessitate increased reliance on energy from nonrenewable sources like fossil fuels [8]. Infrastructure, primarily when optimised for one sector, could inadvertently be detrimental to the other. A system that conserves water might be energy-intensive and vice versa [43]. The economic domain is not exempt from trade-offs. In scenarios where there are economic incentives for conserving one resource, there could inadvertently be increased consumption of the other, creating an imbalance [2]. Furthermore, different regulatory bodies governing the water and energy sectors might sometimes have conflicting directives, posing challenges for seamless DSM. To strike a balance, holistic planning is indispensable. By factoring in both synergies and trade-offs, strategies can be crafted to optimise the energy–water nexus [43]. Engaging stakeholders from both sectors can be invaluable. Through continuous dialogue, challenges can be navigated and opportunities leveraged, creating a symbiotic relationship [38]. Lastly, the linchpin might lie in technological innovation. As we forge ahead, solutions that address trade-offs and bolster synergies will be paramount in realising a sustainable future for both the energy and water sectors [5].

5.7.3 Case studies of integrated approaches to the energy–water nexus

California's prolonged drought conditions significantly impacted water availability and energy production, threatening the state's sustainability [2]. In response, California initiated water conservation methods that simultaneously minimised energy demands. These included wastewater recycling and introducing more efficient irrigation techniques, which ensured a balanced approach to managing the energy–water nexus [1]. As a result, the state observed reduced energy consumption due to these water saving measures, leading to notable environmental and economic benefits [38]. Despite severe water scarcity and relying heavily on imported water, Singapore embarked on an innovative journey to ensure water sustainability [75]. The city-state launched the NEWater initiative, a pioneering project that converts treated wastewater into high-purity water. This water is now used for various purposes, including potable consumption and meeting industrial demands. The initiative reduced Singapore's dependence on external water sources and led to considerable energy savings associated with local water processing, fostering an integrated approach to the energy–water nexus [10].

Morocco, grappling with freshwater scarcity in a growing population, sought renewable energy solutions to meet its water needs [39]. The country developed an innovative approach, establishing solar-powered desalination plants. By harnessing the energy of the sun, these plants efficiently converted seawater into freshwater, addressing the dual challenge of energy and water [62]. The outcome was twofold: a surge in freshwater supply and a significantly reduced carbon footprint due to the reliance on renewable energy [67]. Australia's Murray–Darling Basin, a critical agricultural region, faced challenges due to over-extraction of its water resources, with subsequent rising energy demands [46]. To address these intertwined challenges, the Basin Plan was introduced. This comprehensive strategy integrated water conservation with energy-saving measures, emphasising sustainable agriculture and efficient irrigation practises [71]. The results were remarkable. The region saw improved water efficiency and a corresponding decline in energy consumption, underscoring the benefits of an integrated approach to managing the energy–water nexus [30].

The energy–water nexus is an integrated system that enables the efficient production, utilisation, and management of either one of the resources or both with the support of the other. In such an integrated system, DR could be applied to one section for efficient management while maximising the production of the other resources [76]. Most energy systems, such as wastewater treatment and seawater desalination, required energy for their operations. Therefore, it is more comprehensive to include energy-efficient management in any analysis of the water system [77]. Therefore, most of the applications of DR in water are also considered energy-efficient utilisation [78, 79]. On the other hand, desalination was used in a DR scheme as a deferrable load to mitigate the variable tendency of renewable energy sources in an integrated energy–water system by [80] as an effective energy management approach.

The authors of [81] develop a DR bidding framework that coordinates the operation of RO water desalination plants using load reduction, demand shifting, distributed generation as DR options to minimise operational cost, daily electricity price, renewable energy sources (RES) curtailment, and net load peak shaving. The authors of [82–85] used a ToU-DR programme for the integration of RES with grid connection desalination systems to reduce the impact of carbon emissions and brine production while minimising the cost of freshwater production cost. The results of those studies validate the impact of DR in optimising the energy–water nexus.

In other recent studies, [86, 87] present a DR programme that co-optimises the operation of the RO desalination system and the power system simultaneously. The use of this model has been investigated in the context of a power grid that is abundant in RES. The objective is to reduce curtailment of RES during periods of oversupply by simultaneously optimising the scheduling of the whole system under a centralised entity. The authors of [18] implemented a price-based DR programme in a stochastic two-step multi-objective optimisation method to meet electricity demand, water, cooling, and many other loads in a coordinated cogeneration and energy storage systems. The model shows the flexibility of DR in providing energy efficiency and the overall reliability of such integrated systems.

5.8 Future trends and opportunities

5.8.1 Emerging technologies and innovations in DSM

Increasing resource constraints and mounting environmental concerns underscore the pressing need for advanced DSM technologies to mitigate these challenges [46]. Among the emerging solutions, smart grids and AMI are at the forefront. These modern energy systems enable real-time monitoring of energy consumption, optimising its distribution and enhancing overall efficiency [8]. IoT is also significantly reshaping DSM. In the domestic sphere, devices like smart thermostats and appliances can actively monitor and adjust energy consumption, promoting efficiency [2]. IoT sensors refine energy usage patterns on an industrial scale, promoting optimum operational efficiency [66].

As energy demands fluctuate, storage solutions become paramount. Innovations such as solid-state and flow batteries are emerging, designed to store energy during periods of low demand and distribute it during high-demand phases [59]. Similarly, thermal energy storage, which captures energy as heat or cold, contributes to load levelling and peak shaving [1]. In AI and machine learning (ML), these technologies bring predictive capabilities into DSM. AI-driven models can anticipate energy demand patterns, paving the way for more proactive energy management [10]. Furthermore, ML algorithms can autonomously modulate energy consumption in real-time, responding to external stimuli like renewable energy availability or peak pricing periods [88].

The blockchain is another game changer for energy management. Beyond enabling decentralised energy trading, it offers a transparent and secure mechanism for energy transactions [30]. This decentralisation allows consumers to trade excess energy directly, emphasising community-driven energy solutions. Virtual and augmented reality (VR/AR) are also making their mark on DSM. These technologies offer simulation-based training for DSM professionals, immersing them in real-world scenarios and refining their skills [34]. Additionally, AR apps have the potential to raise consumer awareness by visually representing energy consumption patterns, encouraging sustainable behaviours [59].

Lastly, the emergence of DERs and microgrids signify a shift toward localised energy solutions. Renewable systems, such as solar and wind installations, reduce the burden on central grids [5]. Microgrids, with their advanced management systems, can independently regulate energy distribution, adapting to local demands and ensuring resilience.

5.8.2 Policy recommendations for energy–water nexus DSM

Strategic policy making is crucial for the sustainable management of intertwined water and energy resources [1]. One primary focus should be on improving data collection and integration. By developing integrated data repositories that encompass water and energy metrics, policymakers can make more informed decisions [8]. To this end, the adoption of advanced monitoring technologies, such as IoT devices, is recommended, as they offer real-time insights [2]. In the energy sector, it is imperative to shift towards renewable resources. Encouraging the use of solar and wind power can significantly reduce water consumption compared to traditional energy sources [59]. Furthermore, as hydropower plays a significant role, policies must ensure that it does not adversely impact aquatic ecosystems, striking a balance between energy and water conservation [1].

Innovative approaches in energy production that prioritise water efficiency should be promoted. This includes adopting advanced cooling technologies in power plants, which minimise water usage [10]. Simultaneously, the pairing of renewable energy with desalination processes can enhance both energy efficiency and freshwater production [88]. On the contrary, in the water sector, the emphasis should be on energy efficiency. Upgrading water infrastructure can mitigate energy-draining leakages [30]. Implementing smart water metres can also empower consumers with real-time consumption data, fostering more efficient usage patterns [34].

For holistic nexus management, cross-sectoral collaborations are paramount. Regular dialogues between stakeholders in both sectors can drive coordinated solutions [59]. This integrated approach should be complemented by resource management plans that equally prioritise water and energy, ensuring neither is overlooked [5]. Furthermore, education plays a pivotal role. Governments and organisations should invest in campaigns that educate the public on the intricacies of the energy–water nexus [35]. Concurrently, hosting stakeholder workshops can equip industries with the latest innovations and best practises in this domain.

Regulatory frameworks need to exhibit flexibility. In an ever-evolving world, policies should be equipped to adapt to varying climatic, technological, and socio-economic landscapes [66]. This flexibility should be underpinned by inclusive decision-making processes that take into account diverse stakeholder perspectives [1]. Lastly, economic instruments can be instrumental in driving sustainable consumption. Pricing mechanisms that genuinely reflect the costs of water and energy consumption can incentivise responsible behaviour [8]. Additionally, financial incentives, such as subsidies and grants, can be pivotal in encouraging industries to adopt sustainable technologies and practises [2].

5.8.3 Advancing research and knowledge gaps

Understanding the energy–water nexus within DSM is paramount to achieving sustainability. This nexus is a complex interplay, and research is vital to bridge the existing knowledge gaps [1]. Our current grasp of consumer behaviour reveals that, while many are aware of water and energy usage, not all act on this awareness. Despite consumers expressing concern about sustainability, this only sometimes results to consistent sustainable behaviour [59]. We need comprehensive insights into the motivations, barriers, and potential drivers that could promote sustainable consumer actions. Understanding these dynamics is vital to facilitate better consumer engagement [2].

Various tools and methods exist for consumers to monitor and regulate their energy and water consumption, from household metres to digital applications. However, the need for more user-friendly, intuitive tools, and comprehensive methodologies remains [34]. Various technologies, such as smart metres and water-efficient appliances are available to consumers. However, the potential for innovative solutions in this space is vast. More research is required to develop cutting-edge solutions that simplify consumer conservation [8]. Several campaigns and strategies aim to educate consumers about the intertwined nature of energy and water conservation. However, there needs to be a clear gap in evaluating the real impact of these campaigns. Research could illuminate which strategies resonate most with consumers and why [59].

Numerous economic incentives encourage consumers towards sustainable practises, like rebates on energy-efficient appliances. Yet, it is crucial to assess the actual efficacy of these incentives. Understanding which incentives yield the most significant behavioural shifts and exploring potential new incentives can lead to more refined programmes [5]. In various capacities, existing policies have influenced consumer behaviour in the energy–water nexus. However, an in-depth evaluation of these policies, their effectiveness, and their unintended consequences needs to be improved. Enhanced research can provide insight into how policies can be better framed for desired outcomes [66]. Lastly, while some solutions address energy and water conservation in tandem, there is significant potential for more integrated DSM approaches. The current research landscape offers limited solutions that holistically manage both resources. An emphasis on creating comprehensive solutions, supported by robust research, could revolutionise how we approach DSM in the energy–water nexus.

5.9 Summary

The discourse has revealed several vital insights. There is a noticeable divide between consumer awareness and actions about the energy–water nexus, hinting at a need for deeper exploration into their motivations and barriers. Although tools to monitor consumption are available, they could be enhanced in functionality and user experience. Existing technologies such as smart metres offer avenues for conservation, but there remains ample scope for further innovation. The effectiveness of educational campaigns and economic incentives to promote sustainability merits further evaluation. Additionally, while policies significantly influence consumer behaviour, there is a need to understand their impacts better. Overall, there is significant potential for more integrated approaches to the energy–water nexus within DSM. DSM is pivotal in the energy–water nexus, emphasising the intertwined nature of energy and water conservation and the need for unified approaches. Given the tight link between the two sectors, achievements or challenges in one can ripple into the other, making a comprehensive strategy essential.

The concept of DR represents a dynamic shift in the energy landscape and promises the same in the water system, empowering consumers to engage in resource management actively. By adjusting energy consumption behaviour in response to grid conditions or price signals, the response to demand plays a vital role in achieving a sustainable and efficient energy and water future.

DR in energy management presents an essential and evolving approach to grid optimisation, energy efficiency, and sustainability. While several challenges must be addressed, collective efforts of stakeholders, policymakers, and consumers can unleash the full potential of DR and contribute to building a more resilient and environmentally friendly energy future.

The application of DR principles in water systems offers an innovative approach to optimise water management, enhance water efficiency, and mitigate the impact of water scarcity. By encouraging consumers, industries, and utilities to adjust their water consumption patterns based on real-time data and demand conditions, the response to demand contributes to a more sustainable and resilient water future. Collaboration among stakeholders, supportive policies, and the adoption of smart technologies are crucial for the successful implementation of DR in water systems.

DR applications in energy–water systems offer a promising approach to enhancing resource efficiency, reducing costs, and promoting environmental sustainability. By leveraging DSM strategies, water utilities can optimise their energy consumption while contributing to a more stable and sustainable energy–water nexus. The application of DR in desalination systems particularly offers a promising approach to improve energy efficiency, reduce costs, and promote environmental sustainability. By integrating DR measures with RES and dynamic pricing strategies, desalination operators can optimise their energy consumption while contributing to a more stable and sustainable energy–water nexus. Collaboration between water utilities, energy providers, and policymakers is essential to effectively implement the response to demand in energy–water systems and drive positive environmental and economic results.

For a sustainable energy–water nexus, a multi-faceted strategy is imperative. This encompasses understanding consumer behaviour more deeply, enhancing monitoring tools, driving technological innovations, refining educational campaigns, analysing the impact of economic tools, adjusting policies based on their real-world effects, and emphasising integrated DSM solutions. When these aspects are used in tandem, a sustainable pathway for the energy–water nexus can be charted, safeguarding these vital resources for future generations.

References

[1] Paulet R, Holland P and Morgan D 2021 A meta-review of 10 years of green human resource management: is Green HRM headed towards a roadblock or a revitalisation? *Asia Pacific J. Hum. Resour.* **59** 159–83

[2] Jegatheesan V, Shu L and Jegatheesan L 2021 Producing fit-for-purpose water and recovering resources from various sources: an overview *Environ. Qual. Manag.* **31** 9–28
[3] Pordanjani A H *et al* 2021 Nanofluids: physical phenomena, applications in thermal systems and the environment effects—a critical review *J. Clean. Prod.* **320** 128573
[4] Alaimo L S, Ciommi M, Vardopoulos I, Nosova B and Salvati L 2022 The medium-term impact of the COVID-19 pandemic on population dynamics: the case of Italy *Sustain* **14** 13995
[5] Sahoo M and Sethi N 2022 The dynamic impact of urbanization, structural transformation, and technological innovation on ecological footprint and PM2.5: evidence from newly industrialized countries *Environ. Dev. Sustain.* **24** 4244–77
[6] Pelkmans J 2021 *Nudging China to Go Green: How EU Trade and Cooperation Policies Stimulated China*
[7] Guo X X, Li K L, Liu Y Z, Zhuang M H and Wang C 2022 Toward the economic-environmental sustainability of smallholder farming systems through judicious management strategies and optimized planting structures *Renew. Sustain. Energy Rev.* **165** 112619
[8] Zhou Y and Lund P D 2023 Peer-to-peer energy sharing and trading of renewable energy in smart communities—trading pricing models, decision-making and agent-based collaboration *Renew. Energy* **207** 177–93
[9] Panda S *et al* 2022 Residential demand side management model, optimization and future perspective: a review *Energy Reports* **8** 3727–66
[10] Arora M, Yeow L W, Cheah L and Derrible S 2022 Assessing water circularity in cities: methodological framework with a case study *Resour. Conserv. Recycl.* **178** 106042
[11] Basak S, Dey B and Bhattacharyya B 2022 Demand side management for solving environment constrained economic dispatch of a microgrid system using hybrid MGWOSCACSA algorithm *CAAI Trans. Intell. Technol.* **7** 256–67
[12] Moudgil V, Hewage K, Hussain S A and Sadiq R 2023 Integration of IoT in building energy infrastructure: a critical review on challenges and solutions *Renew. Sustain. Energy Rev.* **174** 113121
[13] Rollert K E 2022 Demand response aggregators as institutional entrepreneurs in the European electricity market *J. Clean. Prod.* **353** 131501
[14] Zhang L, Kuang J, Sun B, Li F and Zhang C 2020 A two-stage operation optimization method of integrated energy systems with demand response and energy storage *Energy* **208** 118423
[15] Hlalele T G, Zhang J, Naidoo R M and Bansal R C 2021 Multi-objective economic dispatch with residential demand response programme under renewable obligation *Energy* **218** 119473
[16] Parvin K *et al* 2022 The future energy internet for utility energy service and demand-side management in smart grid: current practices, challenges and future directions *Sustain. Energy Technol. Assess.* **53** 102648
[17] Ling M and Verma S Enhancing cleaning operations with JSP and servlet-based management systems 2023 *Tensorgate J. Sustain. Tech. and Infrastruct. for Developing Countries* **6** 47–63
[18] Norouzi F, Karimi H and Jadid S 2023 Stochastic electrical, thermal, cooling, water, and hydrogen management of integrated energy systems considering energy storage systems and demand response programs *J. Energy Storage* **72** 108310
[19] Dai J *et al* 2018 Water-energy nexus: a review of methods and tools for macro-assessment *Appl. Energy* **210** 393–408

[20] Pezhmani Y, Oskouei M Z, Rezaei N and Mehrjerdi H 2022 A centralized stochastic optimal dispatching strategy of networked multi-carrier microgrids considering transactive energy and integrated demand response: application to water–energy nexus *Sustain. Energy, Grids Networks* **31** 100751
[21] Xiang Y, Cai H, Gu C and Shen X 2020 Cost-benefit analysis of integrated energy system planning considering demand response *Energy* **192** 116632
[22] Nirbheram J S, Mahesh A and Bhimaraju A 2023 Techno-economic analysis of grid-connected hybrid renewable energy system adapting hybrid demand response program and novel energy management strategy *Renew. Energy* **212** 1–16
[23] Mohajeryami S, Moghaddam I N, Doostan M, Vatani B and Schwarz P 2016 A novel economic model for price-based demand response *Electr. Power Syst. Res.* **135** 1–9
[24] Zhang F, de Dear R and Candido C 2016 Thermal comfort during temperature cycles induced by direct load control strategies of peak electricity demand management *Build. Environ.* **103** 9–20
[25] Cui Q, Wang X, Wang X and Zhang Y 2016 Residential appliances direct load control in real-time using cooperative game *IEEE Trans. Power Syst.* **31** 226–33
[26] Dadkhah A, Bayati N, Shafie-khah M, Vandevelde L and Catalão J P S 2022 Optimal price-based and emergency demand response programs considering consumers preferences *Int. J. Electr. Power Energy Syst.* **138** 107890
[27] Alipour M, Zare K, Seyedi H and Jalali M 2019 Real-time price-based demand response model for combined heat and power systems *Energy* **168** 1119–27
[28] Bagriyanik F G, Sonmez M A and Bagriyanik M 2023 The impact of mutual interaction between electric and water distribution systems via demand response *Sustain. Energy, Grids Networks* **34** 101008
[29] Saeian H, Niknam T, Zare M and Aghaei J 2022 Coordinated optimal bidding strategies methods of aggregated microgrids: a game theory-based demand side management under an electricity market environment *Energy* **245** 123205
[30] Otoum S, Ridhawi I A and Mouftah H 2023 A federated learning and blockchain-enabled sustainable energy trade at the edge: a framework for Industry 4.0 *IEEE Internet Things J.* **10** 3018–26
[31] Wen M, Zhou C and Konstantin M 2023 Deep neural network for predicting changing market demands in the energy sector for a sustainable economy *Energies* **16** 2407
[32] Scarabaggio P, Grammatico S, Carli R and Dotoli M 2022 Distributed demand side management with stochastic wind power forecasting *IEEE Trans. Control Syst. Technol.* **30** 97–112
[33] Anselmo S, Boccardo P, Cimadom A, Corgnati S P and La Riccia L 2022 *2.5d maps supporting energy policy in Turin* Politecnico di Torino
[34] Cierco Jimenez R *et al* 2022 Machine learning computational tools to assist the performance of systematic reviews: a mapping review *BMC Med. Res. Methodol.* **22** 1–14
[35] Khaleel M 2023 Intelligent control techniques for microgrid systems *Brill. Res. Artif. Intell.* **3** 56–67
[36] Usman R, Mirzania P, Alnaser S W, Hart P and Long C 2022 Systematic review of demand-side management strategies in power systems of developed and developing countries *Energies* **15** 7858
[37] Kanakadhurga D and Prabaharan N 2022 Demand side management in microgrid: a critical review of key issues and recent trends *Renew. Sustain. Energy Rev.* **156** 111915

[38] Rahman M M, Khan I, Field D L, Techato K and Alameh K 2022 Powering agriculture: present status, future potential, and challenges of renewable energy applications *Renew. Energy* **188** 731–49

[39] 2022 *Food, Energy, and Water Nexus* ed C Ray, S Muddu and S Sharma (Cham: Springer)

[40] Almohaimeed S A, Suryanarayanan S and O'Neill P 2021 Reducing carbon dioxide emissions from electricity sector using demand side management *Energy Sources, Part A Recover. Util. Environ. Eff.* 1–21

[41] Levine A and Ghezzi M 2022 Pedagogical strategies for teaching the DSM: centering diversity and equity *Adv. Soc. Work* **22** 133–44

[42] Thakur T *et al* 2021 Smart water conservation through a machine learning and blockchain-enabled decentralized edge computing network *Appl. Soft Comput.* **106** 107274

[43] Peng J *et al* 2023 The conflicts of agricultural water supply and demand under climate change in a typical arid land watershed of Central Asia *J. Hydrol. Reg. Stud.* **47** 101384

[44] Guelpa E and Verda V 2021 Demand response and other demand side management techniques for district heating: a review *Energy* **219** 119440

[45] Gyamfi S, Diawuo F A, Asuamah E Y and Effah E 2022 The role of demand-side management in sustainable energy sector development *Renewable Electricity and Sustainability: Prospects in Developing Economies* (New York: Elsevier)

[46] Penny J, Ordens C M, Barnett S, Djordjević S and Chen A S 2023 Small-scale land use change modelling using transient groundwater levels and salinities as driving factors—An example from a sub-catchment of Australia's Murray-Darling Basin *Agric. Water Manag.* **278** 108174

[47] Emaminejad S A, Sparks J and Cusick R D 2023 Integrating bio-electrochemical sensors and machine learning to predict the efficacy of biological nutrient removal processes at water resource recovery facilities *Environ. Sci. Technol.* **57** 18372–81

[48] Tan Q, Liu Y and Zhang X 2021 Stochastic optimization framework of the energy–water-emissions nexus for regional power system planning considering multiple uncertainty *J. Clean. Prod.* **281** 124470

[49] Iqbal S *et al* 2021 A comprehensive review on residential demand side management strategies in smart grid environment *Sustain* **13** 7170

[50] Norbu S, Couraud B, Robu V, Andoni M and Flynn D 2021 Modelling the redistribution of benefits from joint investments in community energy projects *Appl. Energy* **287** 116575

[51] Groppi D, Pfeifer A, Garcia D A, Krajačić G and Duić N 2021 A review on energy storage and demand side management solutions in smart energy islands *Renew. Sustain. Energy Rev.* **135** 110183

[52] Prajapati V K and Mahajan V 2021 Reliability assessment and congestion management of power system with energy storage system and uncertain renewable resources *Energy* **215** 119134

[53] Ihsan A, Jeppesen M and Brear M J 2019 Impact of demand response on the optimal, techno-economic performance of a hybrid, renewable energy power plant *Appl. Energy* **238** 972–84

[54] Nwulu N I and Xia X 2017 Optimal dispatch for a microgrid incorporating renewables and demand response *Renew. Energy* **101** 16–28

[55] Li J *et al* 2023 Optimal planning of electricity–hydrogen hybrid energy storage system considering demand response in active distribution network *Energy* **273** 127142

[56] Luo X, Liu Y, Feng P, Gao Y and Guo Z 2021 Optimization of a solar-based integrated energy system considering interaction between generation, network, and demand side *Appl. Energy* **294** 116931
[57] Muttaqi K M, Aghaei J, Ganapathy V and Nezhad A E 2015 Technical challenges for electric power industries with implementation of distribution system automation in smart grids *Renew. Sustain. Energy Rev.* **46** 129–42
[58] Jordehi A R 2019 Optimisation of demand response in electric power systems, a review *Renew. Sustain. Energy Rev.* **103** 308–19
[59] Metzger P, Mendonça S, Silva J A and Damásio B 2023 Battery innovation and the circular economy: what are patents revealing? *Renew. Energy* **209** 516–32
[60] Kaluarachchi Y 2022 Implementing data-driven smart city applications for future cities *Smart Cities* **5** 455–74
[61] Wang X C, Jiang P, Yang L, Fan Y V, Klemeš J J and Wang Y 2021 Extended water-energy nexus contribution to environmentally-related sustainable development goals *Renew. Sustain. Energy Rev.* **150** 111485
[62] Al-Obaidi M A, Zubo R H A, Rashid F L, Dakkama H J, Abd-Alhameed R and Mujtaba I M 2022 Evaluation of solar energy powered seawater desalination processes: a review *Energies* **15** 6562
[63] Dicce R P and Ewers M C 2021 Solar labor market transitions in the United Arab Emirates *Geoforum* **124** 54–64
[64] de Sá Silva A C R, Bimbato A M, Balestieri J A P and Vilanova M R N 2022 Exploring environmental, economic and social aspects of rainwater harvesting systems: a review *Sustain. Cities Soc.* **76** 103475
[65] Ingrao C, Strippoli R, Lagioia G and Huisingh D 2023 Water scarcity in agriculture: an overview of causes, impacts and approaches for reducing the risks *Heliyon* **9** e18507
[66] Toomey A H 2023 Why facts don't change minds: insights from cognitive science for the improved communication of conservation research *Biol. Conserv.* **278** 109886
[67] Usman A, Ozturk I, Naqvi S M M A, Ullah S and Javed M I 2022 Revealing the nexus between nuclear energy and ecological footprint in STIRPAT model of advanced economies: fresh evidence from novel CS-ARDL model *Prog. Nucl. Energy* **148** 104220
[68] Robinson B L, Clifford M J and Selby G 2023 Towards fair, just and equitable energy ecosystems through smart monitoring of household-scale biogas plants in Kenya *Energy Res. Soc. Sci.* **98** 103007
[69] Hong J, Lee W, Ha Kim J, Kim J, Park I and Har D 2016 Smart water grid: desalination water management platform *Desalin. Water Treat.* **57** 2845–54
[70] Fu H, Niu J, Wu Z, Cheng B, Guo X and Zuo J 2022 Exploration of public stereotypes of supply-and-demand characteristics of recycled water infrastructure—evidence from an event-related potential experiment in Xi'an, China *J. Environ. Manage.* **322** 116103
[71] Zhang Y *et al* 2023 Agricultural drought characteristics in a typical plain region considering irrigation, crop growth, and water demand impacts *Agric. Water Manag.* **282** 108266
[72] Addo I B, Thoms M C and Parsons M 2018 Household water use and conservation behavior: a meta-analysis *Water Resour. Res.* **54** 8381–400
[73] Mkireb C, Dembélé A, Jouglet A and Denoeux T 2019 Robust optimization of demand response power bids for drinking water systems *Appl. Energy* **238** 1036–47
[74] Jaafar J *et al* 2018 Forward osmosis for desalination application *Membrane Separation Principles and Applications: From Material Selection to Mechanisms and Industrial Uses*

ed A F Ismail, M A Rahman, M H D Othman and T Matsuura (New York: Elsevier) pp 315–37

[75] Zhao H, Miller T R, Ishii N and Kawasaki A 2022 Global spatio-temporal change assessment in interregional water stress footprint in China by a high resolution MRIO model *Sci. Total Environ.* **841** 156682

[76] Ghasemi-Marzbali A, Shafiei M and Ahmadiahangar R 2023 Day-ahead economical planning of multi-vector energy district considering demand response program *Appl. Energy* **332** 120351

[77] Okampo E J and Nwulu N 2021 Optimisation of renewable energy powered reverse osmosis desalination systems: a state-of-the-art review *Renew. Sustain. Energy Rev.* **140** 110712

[78] Li Q, Yu S, Al-Sumaiti A and Turitsyn K 2018 Micro water-energy nexus: optimal demand-side management and quasi-convex hull relaxation *IEEE Trans. Control Netw. Syst.* **6** 1313–22

[79] Okampo E J and Nwulu N I 2020 Optimal energy mix for a reverse osmosis desalination unit considering demand response osmosis *J. Eng. Des. Technol.* **18** 1287–303

[80] Bognar K, Blechinger P and Behrendt F 2012 Seawater desalination in micro grids: an integrated planning approach *Energy. Sustain. Soc.* **2** 14

[81] Elsir M, Al-Sumaiti A S, El Moursi M S and Al-Awami A T 2023 Coordinating the day-ahead operation scheduling for demand response and water desalination plants in smart grid *Appl. Energy* **335** 120770

[82] Okampo E J and Nwulu N 2020 Optimal design and techno-economic evaluation of renewable energy powered combined reverse osmosis desalination and brine treatment unit *Desalin. Water Treat.* **202** 27–37

[83] Okampo E J and Nwulu N 2022 Economic and reliability assessment of hybrid PRO-RO desalination systems using brine for salinity gradient energy production *Sustainability* **14** 3328

[84] Okampo E J and Nwulu N 2021 Techno-economic evaluation of reverse osmosis desalination system considering emission cost and demand response *Sustain. Energy Technol. Assess.* **46** 101252

[85] Giudici F, Castelletti A, Garofalo E, Giuliani M and Maier H R 2019 Dynamic, multi-objective optimal design and operation of water-energy systems for small, off-grid islands *Appl. Energy* **250** 605–16

[86] Elsir M, Al-Awami A T, Antar M A, Oikonomou K and Parvania M 2023 Risk-based operation coordination of water desalination and renewable-rich power systems *IEEE Trans. Power Syst.* **38** 1162–75

[87] Elsir M and Al-Awami A T 2020 Coordinating the operation of water desalination and a renewables-rich electric power grid *2020 IEEE Power & Energy Society General Meeting*

[88] Devlin A and Yang A 2022 Regional supply chains for decarbonising steel: energy efficiency and green premium mitigation *Energy Convers. Manag.* **254** 115268

Chapter 6

Techno-economic evaluation of desalination systems

The increasing global demand for freshwater, coupled with the challenges of water scarcity and population growth, has propelled desalination to the forefront as a vital solution. However, it is still widely considered expensive. The reverse osmosis method, which is widely used because of it cost-effectiveness, still requires high energy consumption. Conventional fossil energy sources are known to release carbon emissions, which have significant environmental and economic consequences, making renewable energy sources (RES) more efficient and preferred options. However, most RES face limitations due to their intermittent nature. This chapter presents mathematical optimisation models that compare the utilisation of a standard grid and diesel generator system with a system that incorporates RES to power a reverse osmosis (RO) desalination plant. The formulation of the model incorporates a balance between the supply and demand of electricity, the demand for water and its production, cost functions, and the implementation of a time-of-use demand response programme. The cost functions encompass the carbon emission cost, demand response cost, and components cost. The objectives of the optimisation models are to reduce the annualised cost of the system and carbon emissions, while simultaneously optimising the quantity of freshwater production. Two case studies, each with three scenarios of optimal sizing of energy mix for an RO unit were evaluated using the levelized cost of energy (LCOE), cost of water (COW), and annualised cost of the system (ACS) matrices, and subsequently, the outcomes were compared. The findings show the influence of RES and demand response programmes on minimising the overall COW desalination.

doi:10.1088/978-0-7503-6154-5ch6

6.1 Introduction

The concept of techno-economic assessment of desalination systems allows for a systematic examination of the construction process of such systems. This is done mostly through design models, which consider the optimal cost implications amid the technical constraints of the system. In other words, optimal techno-economic assessment evaluates a desalination system from both technical and economic perspectives, to critically identify and minimise costs while maintaining or improving the performance of different technology factors. As discussed earlier in the previous chapter, optimal sizing and operational optimisation are considered fundamental for techno-economic evaluation of desalination systems. These optimisation approaches could be multi-objective, considering the different parameters, variables, and constraints of the system. In this chapter, we present optimisation models that focus on key factors like the sizing of different energy sources, the impact of carbon emission cost, and the time-of-use demand response programme on the overall cost of an RO desalination system.

Optimal sizing of energy sources for a water treatment plant involves selecting the most suitable energy source for a specific treatment method considering cost-effectiveness and within the balance of technical constraints. The need for clean, reliable, and affordable sources of energy is as paramount as the need for freshwater itself. The effects of climate change and population growth have made these needs more pressing and threaten livelihood security and developmental sustainability. For so long, in the history of man through the industrial revolution to the present day civilisation, there has been an overdependence on fossil fuel sources of energy. This has made these forms of energy sources the traditional and most conventional energy sources in use, even for water treatment systems such as desalination. The reason for this dependence on traditional energy sources may be the easy access and simple process of conversion from their original form to another energy form needed for utilities. Additionally, conventional energy sources are considered relatively cheaper to maintain until factors such as the environmental effects of carbon emissions are considered. Carbon emissions have both environmental and economic implications, such as the carbon tax implemented as a regulation in different countries. The challenges of carbon footprints, which are referred to as global warming impacts (GWIs), and their economic implication shed more light on the importance of the integration of renewable energy in the energy mix for desalination. However, the intermittent nature of RES and their cost means it might not be wise to outright replace traditional energy sources with RES. Therefore, an integrated energy mix of conventional and RES models with a time-of-use (ToU) demand response (DR) programme is expected to be more cost-effective, reliable, and environmentally sustainable to power a desalination system [1, 2].

Demand-side management is a system designed to minimise energy consumption while improving the reliability of local electricity grid either by consumer incentive or ToU response programme. Many utilities use demand-side management to identify and influence energy consumption. The aim of this energy usage management programme is to reduce the cost of energy consumption and environmental impact. The implementation of this DR programme alongside desalination systems powered by different energy mixes will help to determine the most cost-effective schedule for the energy generating units.

The contribution of this chapter is therefore to present two case studies of optimisation models that can help to determine the optimal cost of freshwater production considering different energy supply sources while optimising the production capacity of RO units for maximum freshwater production, and furthermore, the impact of carbon emission and demand-side management on the cost of RO desalination systems.

6.2 The effect of DR on the energy mix of the desalination system

In general, the optimal sizing of energy sources is performed with regard to the peak demand, which are usually a short period in the daily load curves. Therefore, to avoid over-using the intended renewable energy source that is intended to be integrated with grid electricity to meet the peak demand, a DR strategy is employed. The DR programme is used to adjust consumer demand patterns in line with changes in electricity prices over time. It mostly requires shifting or reducing the demand from when electricity prices are higher to when prices are lower. Therefore, an electricity price structure is needed to carry out a DR programme. The electricity price structure is a policy framework for government and utilities, subject to location, and it is country specific. It is intended to improve the economic benefit of consumers while simultaneously modifying their consumption behaviour. As a result, the DR programme is classified as incentive-based consumer or ToU price-based. In this book, we shall be implementing the ToU price-based DR programme. Since the electricity pricing structure is location or country and utility specific, the Eskom electricity price structure for South Africa is implemented in the optimal desalination model presented in subsequent sections and chapters.

6.2.1 Energy supply and demand modelling

Seawater desalination is energy intensive. The amount of energy needed for a desalination unit to produce water depends on the desalination method, plant design, temperature and salinity of the feed water [3]. Specific energy consumption (SEC) in kWh m^{-3} is the energy needed to desalinate 1 m^3 of freshwater. It is different for every method of desalination. For example, multi-effect distillation requires between 14.45 and 21.35 kWh m^{-3} to desalinate 1 m^3 of freshwater, while multi-stage flash distillation (MSF) needs between 19.58 and 27.25 kWh m^{-3}, and mechanical vapour compression uses between 7 and 12 kWh m^{-3}. Furthermore, thermal vapour compression (TVC) requires about 16.3 kWh m^{-3}, and electro-dialysis consumes between 0.7 and 5.5 kWh m^{-3} [3, 4] while RO uses 2–4 kWh to produce 1 m^3 of freshwater [5, 6]. RO is the most widely used desalination method—its technology is advanced and easy to integrate with energy sources and other desalination methods of desalination [7, 8]. Furthermore, it is relatively lower than most thermal desalination methods. Hence, this techno-economic assignment will focus on evaluating RO desalination systems. Therefore, the hourly power requirement of the RO desalination unit is the main demand that must be satisfied by the energy supply mix of grid electricity and diesel generators, both considered traditional fossil fuel energy sources integrated with a renewable energy source of either photovoltaic or wind energy.

The following paragraphs of this section detail the sequence of the optimisation model with the components of each energy source and the RO system represented in a mathematical formulation. This is such that the economic impact of the components and different sections of the entire system is evaluated using an ACS matrix to determine the unit COW.

The power supply from the grid acts as a backup to supplement the erratic supply from photovoltaic (PV) energy to meet the power requirement of the desalination unit. This ensures that the unit continues to function normally. When the amount of electricity generated by PV systems is more than the amount of energy required to meet load demand, the surplus power is sold to the grid. Therefore, the hourly power that is imported from the grid, denoted by the symbol $P_{gi}(t)$, and the hourly power that is exported to the grid, denoted by the symbol $P_{ge}(t)$, are both optimised variables that range from zero to the RO unit's maximum hourly load demand. These quantities are represented by the symbols (6.1) and (6.2), respectively

$$0 \leqslant P_{gi}(t) \leqslant P_{gi}{}^{max} \tag{6.1}$$

$$0 \leqslant P_{ge}(t) \leqslant P_{ge}{}^{max}. \tag{6.2}$$

The maximum transferable power to and from the grid is assumed to be equal:

$$P_{gi}{}^{max} = P_{ge}{}^{max}. \tag{6.3}$$

The difference between the amount of power that is imported and the amount that is exported to the grid is one of the factors that determines the cost of grid electricity. It is assumed that the unit cost of obtaining electricity is equivalent to the unit cost of selling power back to the grid. Therefore, the extra energy output not required by the desalination unit is sold back to the grid to pay for the cost of imported electricity or, at the very least, minimise the importation cost of energy from the grid.

$$\mathrm{CP}_{gi}(t) = P_{gi}(t) \times \mathrm{PR}(t) \tag{6.4}$$

$$\mathrm{CP}_{ge}(t) = P_{ge}(t) \times \mathrm{PR}(t) \tag{6.5}$$

$$\mathrm{CP}_{g} = \sum (\mathrm{CP}_{gi}(t) \times \mathrm{PR}(t) - P_{ge}(t) \times \mathrm{PR}(t)) \tag{6.6}$$

$$\forall\, t = 1, 2, 3, \ldots 8760$$

where:

$PR(t)$ denotes the hourly unit price of transferable grid power, whereas CP_g is the total annual cost of transferable grid power.

The diesel generator (DG) is used in this study as a second backup to make up for the inconsistency of the RES. Its output power ($P_{dg}(t)$) is an optimised variable that is a function of its maximum capacity, as expressed in (6.7)

$$P_{dg}{}^{min} \leqslant P_{dg}(t) \leqslant P_{dg}{}^{max}. \tag{6.7}$$

The hourly fuel consumption of the DG, $FC\ (t)\ (Lh^{-1})$, is a function of its electrical power output [9].

$$FC(t) = a \times P_{dg}(t) + b \times P_{dg}{}^{N} \tag{6.8}$$

where:

$a = 0.2461\ \mathrm{kW^{-1}h^{-1}}$ is the slope of the fuel curve and $b = 0.081\ 45\ \mathrm{l\ kW^{-1}h^{-1}}$ is the intercept coefficient of the rated fuel curve with $P_{dg}{}^{N}$ being the generator nominal power of the generator [10, 11]. Therefore, the fuel cost of the DG is a function of the unit price of fuel (Fpr), which in this study is 1 \$ L^{-1}, and the amount of fuel consumed to produce its output power. Hence, the total annual fuel cost is expressed as (6.9) [10, 11]

$$\mathrm{FC_{DG}} = \sum_{t=1}^{8760} \mathrm{FC}(t) \times \mathrm{Fpr}. \tag{6.9}$$

The initial capital cost of the DG (IC_{DG}) and the maintenance cost (AMC_{DG}) are \$7495 and 0.3 \$ h^{-1}, respectively [11].

Furthermore, the energy mix includes renewable energy of PV, which has to be adequately sized for optimal performance. Therefore, the hourly output power supply by the PV array is given as [12, 13]

$$P_{\mathrm{PV}}(t) = \mathrm{APV} \times \acute{\eta} \times \mathrm{SI}(t) \tag{6.10}$$

where APV is the area of the PV array in m^2, $SI(t)$ is the hourly solar irradiation, which is location-based, and $\acute{\eta}$ is the PV efficiency.

The evaluation of the PV cost is based on the area of the PV array, the initial capital cost, and the maintenance cost throughout its useful life, as expressed in (6.11) and (6.12) [11, 14]

$$\mathrm{IC_{PV}} = \mathrm{APV} \times \mathrm{C_{PV}} \tag{6.11}$$

$$\mathrm{AMC_{PV}} = \mathrm{APV} \times \mathrm{MC_{PV}}. \tag{6.12}$$

6.2.2 Desalination plant energy demand

The power demand $P_{WD}(t)$ of the RO desalination unit is the load that is required to be satisfied and depends on the *SEC* to produce 1 m^3 of freshwater; this value ranges from 3 to 4 kWh m^{-3} for RO [10] and according to [6, 11, 13, 15] it currently ranges from 2–5 kWh m^{-3}, which in this study is 3 kwh m^{-3} and the actual volume of water ($QW(t)$) produced per hour [5, 16]. Thus, to model this mathematically, we have:

$$P_{\mathrm{WD}}(t) = \mathrm{QW}(t) \times \mathrm{SEC}. \tag{6.13}$$

The daily water production capacity can be represented as

$$\mathrm{DQW} = \sum_{t=1}^{24} \mathrm{QW}(t). \tag{6.14}$$

The water distribution model and network are not considered in this study. Therefore, the capacity of the water tank (W_{TK}) expressed in m^3 is assumed to be twice the daily capacity of water production to make enough space available for excess water production and in the event of water remaining from the previous day

$$W_{TK} = \mathrm{DQW} \times 2. \tag{6.15}$$

The RO desalination cost model includes the initial capital cost (IC_{RO}), annual maintenance and operational cost (AMC_{RO}), annual membrane replacement cost (AC_{MR}), annual treatment chemical cost (AC_{CH}), and water tank cost (CW_{TK}) as adopted by references [10, 11]

$$IC_{RO} = C_{RO} \times \mathrm{DQW} \tag{6.16}$$

$$ICW_{TK} = CW_{TK} \times W_{TK}. \tag{6.17}$$

6.2.3 DR model

To maintain a stable power flow in the designed system, it is assumed that the maximum allowable demand variation (increase/decrease) per hour is 30% of total demand at that hour, as represented in (5.18). Therefore, the DR load, which is the load shifted as a result of the response, is an optimised variable, as expressed in (6.19) [17, 18]. This change in demand load lies between the minimum allowable load cut and the maximum allowable load addition at any hour of the day for optimal load balancing. Thus:

$$\Delta L^{\max}(t) = 0.3 \times P_{WD}(t) \tag{6.18}$$

and

$$-\Delta L^{\max}(t) \leqslant \Delta L(t) \leqslant \Delta L^{\max}(t). \tag{6.19}$$

The application of ToU DR optimally shifts loads from peak hours when the price of demand is highest to standard or off-peak hours, yet allows the demand at every hour to be met. The cost of DR is the difference between the cost of power demand before and after ToU DR. It is expressed as [19]:

$$\Delta \mathrm{LC} = \sum [P_{WD}(t) \times \lambda(t) - (P_{WD}(t) - \Delta L(t)) \times \lambda(t)]. \tag{6.20}$$

In this study, the annualised cost matrix is used for the economic evaluation of the RO desalination system powered by the different energy sources to determine the cost of freshwater and LCOE. ACS involves the capital recovery factor (CRF) and the total cost of the system, which includes the total initial capital cost (TICC), cost of grid power (CGP) and the total maintenance and operational cost (TMC) of all components that make up the system [20, 21, 22]

$$\mathrm{TICC} = (\mathrm{IC_{PV}} + \mathrm{IC_{DG}} + \mathrm{IC_{RO}} + \mathrm{ICW_{TK}}) \times \mathrm{dr} \tag{6.21}$$

where,

$\mathrm{dr} = \frac{1}{(1+i)^{20}}$, the discount rate for 20 years' lifetime of the system

$$\mathrm{TMC} = \mathrm{AMC_{PV}} + \mathrm{AMC_{DG}} + \mathrm{AMC_{RO}} \tag{6.22}$$

$$\mathrm{ACS} = \mathrm{TICC} \times \mathrm{CRF} + \mathrm{TMC} + \mathrm{AC_{MR}} \times \mathrm{CRF} + \mathrm{AC_{CH}} + \mathrm{CP_G} + \mathrm{FC_{DG}} + \Delta\mathrm{LC} \tag{6.23}$$

where,

$$\mathrm{CRF} = \frac{i(1+i)^n}{(1+i)^n - 1} \qquad n = 1.\ldots 19 \tag{6.24}$$

$$\mathrm{COW} = \mathrm{ACS}/\sum \mathrm{QW}(t) \tag{6.25}$$

$$\mathrm{LCOE} = \frac{\mathrm{ACS} + P_{\mathrm{WD}}(t)}{\sum P_{\mathrm{PV}}(t) + Pg(t) + GP_i(t)} \tag{6.26}$$

$$\forall\, t = 1, 2, 3, \ldots 8760.$$

6.2.4 Optimisation model

The optimisation problem is formulated as a multi-objective linear programming problem as expressed in (6.27), with the objective of minimising ACS, which includes total initial capital cost, membrane replacement cost, maintenance cost, treatment chemical cost of the RO unit, grid transferable power cost, DG fuel cost, and DR cost, subject to constraints expressed by (6.28–6.32). This multi-optimisation problem is solved using the CPLEX solver of AIMMS.

Objective Function

Min ACS =

$$\min \sum (\mathrm{TICC} + \mathrm{AC_{MR}})\mathrm{CRF} + \mathrm{TMC} + \mathrm{AC_{CH}} + \mathrm{CGP} + \mathrm{FC_{DG}} + \Delta\mathrm{LC} \tag{6.27}$$

S.t.

$$P_{\mathrm{PV}}(t) + P_g(t) + \mathrm{GP}_i(t) = P_{\mathrm{WD}}(t) + \Delta L(t) + \mathrm{GP}_e(t) \tag{6.28}$$

$$\mathrm{QW}(t) = \mathrm{WD}(t) \tag{6.29}$$

$$\mathrm{QW}^{\min} \leqslant \mathrm{QW}(t) \leqslant \mathrm{QW}^{\max} \tag{6.30}$$

$$P_{\mathrm{WD}}{}^{\min} \leqslant P_{\mathrm{WD}}(t) \leqslant P_{\mathrm{WD}}{}^{\max} \tag{6.31}$$

$$\forall\, t = 1, 2, 3, \ldots 8760$$

$$\mathrm{APV} \geqslant 0. \tag{6.32}$$

Constraint (6.28) expresses the power balance that ensures the power supply from PVs, the DG, and the grid at any time *t*, equalling the total demand by the RO desalination unit and the power exported to the grid at the same time *t*. Constraint (6.29) is the water balance that ensures that the water produced at any hour *t* equals the water demand at that hour *t*, or exceeds it, while constraint (6.30) ensures water produced per hour remains within the required limits. Constraint (6.31) is the limit of power required to produce water at any time *t*, and constraint (6.32) expresses the limit of the area of PVs. Other constraints include (6.1), (6.2), (6.3), and (6.19), which express the limit of grid-imported power, limit of grid exported power, maximum allowable transfer power, and limit of DR load, respectively. Figure 6.1 represents the model flow chart.

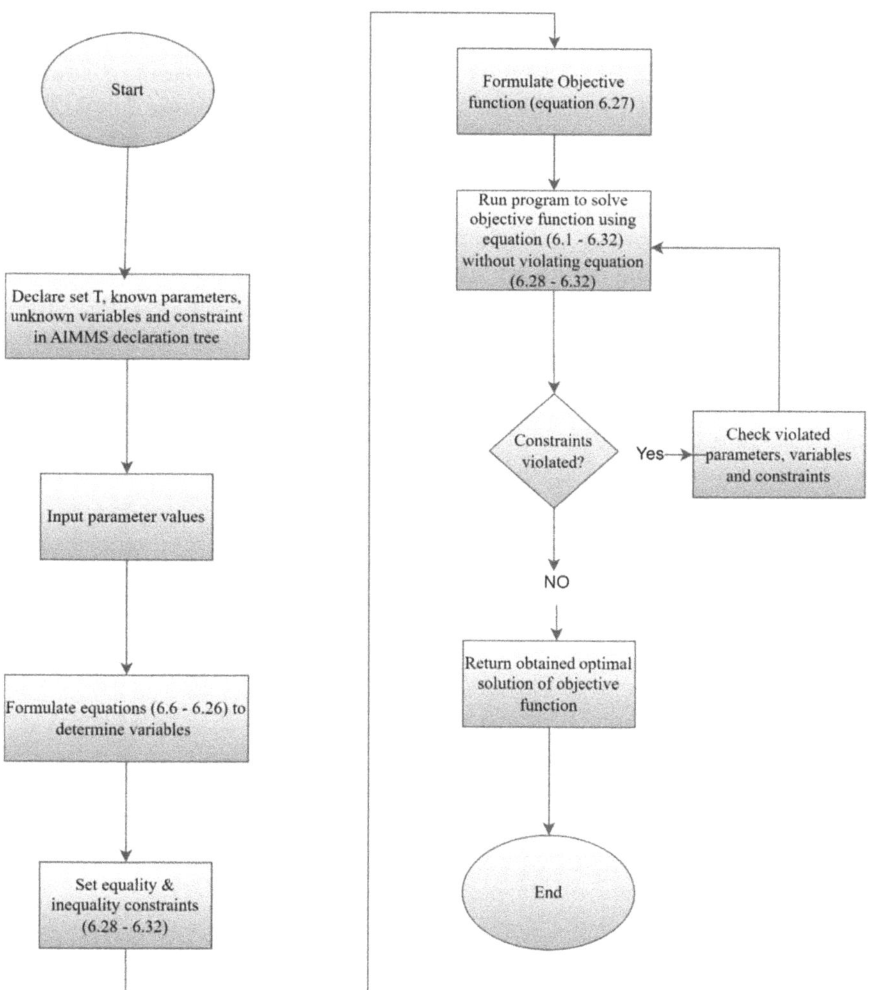

Figure 6.1. Model flow chart.

6.2.5 Case study

To demonstrate the operation of this optimisation model, this section presents three different scenarios to evaluate the RO desalination system. In scenario 1, an RO desalination system is powered by the grid and the DG in the case of using traditional energy sources. The model for this scenario does not include any RES in the energy mix and does not apply a DR programme on the demand-side management. The second scenario incorporates a PV, renewable energy source with the traditional energy sources (grid and DG), but also left out a DR programme on the demand management side. Lastly, scenario 3 is an RO desalination system powered by the grid, a DG, and PVs, alongside incorporating a ToU DR programme on the demand side.

Metrological data were gathered from the Southern African Universities Radiometric Network, namely Stellenbosch University in South Africa's Western Cape province. For example, figure 6.2 represents an average day of solar radiation of the areas under study. Also, the ToU energy price (in US dollars) for South Africa obtained from the Eskom schedule of standard prices for Eskom tariffs 2019/2020 [23] is used for the DR programme. Figure 6.3 represents the ToU price per hour of a day.

Innovation in RO desalination technology is the ideal for South Africa, according to [24]. The Western Cape region of South Africa was chosen for this study due to its proximity to the shoreline of the Atlantic Ocean and the area's current freshwater shortage in the area [1]. The water on the West Coast is often cold and salty. As a result, high operating pressure, low membrane fluxes, and, hence, high energy are required [24]. Due to these issues, desalination is highly expensive, although it is still the preferred option for freshwater sustainability [1]. South Africa already has various desalination facilities in operation, the largest of which is located in Mossel Bay and has a capacity of 15 Ml/day [25]. Regardless of the few operational

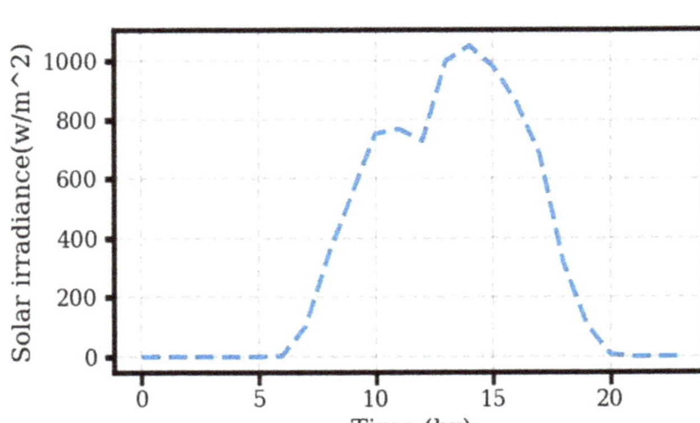

Figure 6.2. Average day of solar irradiance of Cape Town, South Africa [1, 2]. Reprinted from [2], copyright (2021) with permission from Elsevier.

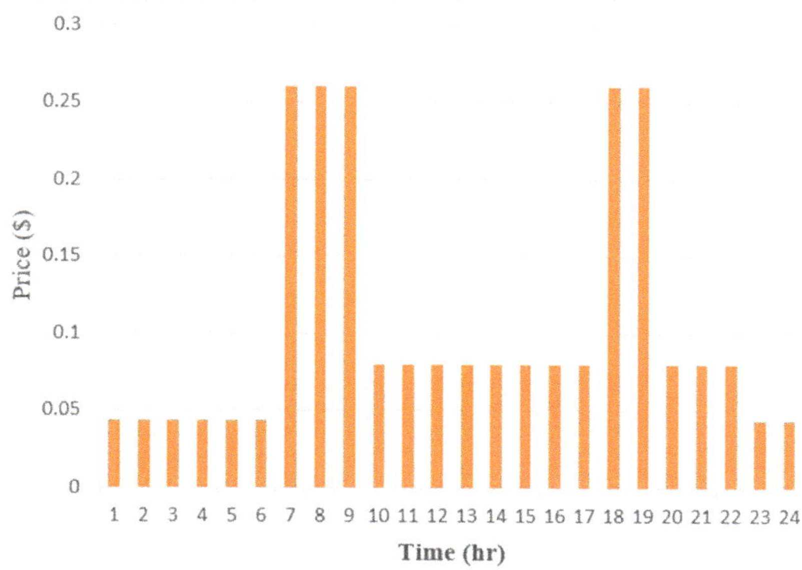

Figure 6.3. Hourly price of the grid power from the South African utility company [1, 2]. Reprinted from [2], copyright (2021) with permission from Elsevier.

desalination plants, recent investigations have revealed that the area still needs approximately 150 Ml/day of freshwater to meet demand and a consistent increase of more than 30 Ml per year to achieve approximately 350 Ml/day to ensure sustainability in the near future [26]. Although a more permanent seawater desalination plant has been proposed with a capacity of more than 150 Ml/day, the capacity of the desalination unit proposed in this investigation is estimated to be 1250–1800 m^3/day, depending on the predicted daily water use. The production of desalinated water is not scaled up or down; it is a continuous daily production rate. Figure 6.4 [1] shows the assumed hourly water demand curve based on the expected social water usage at different times of the day. Table 6.1 shows other data used in this study, particularly for component cost assessment.

6.2.6 Results and discussion

The hourly electricity supply from the grid and the DG is shown in figure 6.5. To provide a clear contrast, an average daily simulation is utilised. Figure 6.6 displays the hourly power production and extra power exported to the grid from scenario 2's energy supply mix. Figure 6.7(a) shows the hourly power supply mix for scenario 3, while figure 6.7(b) shows the change in the load curve before and after the DR programme was applied. Figure 6.8 shows the amount of water delivered in each situation. While table 6.1 summarises the results of daily power supply from the various mix of energy sources, table 6.2 summarises the outcomes of optimised parameters.

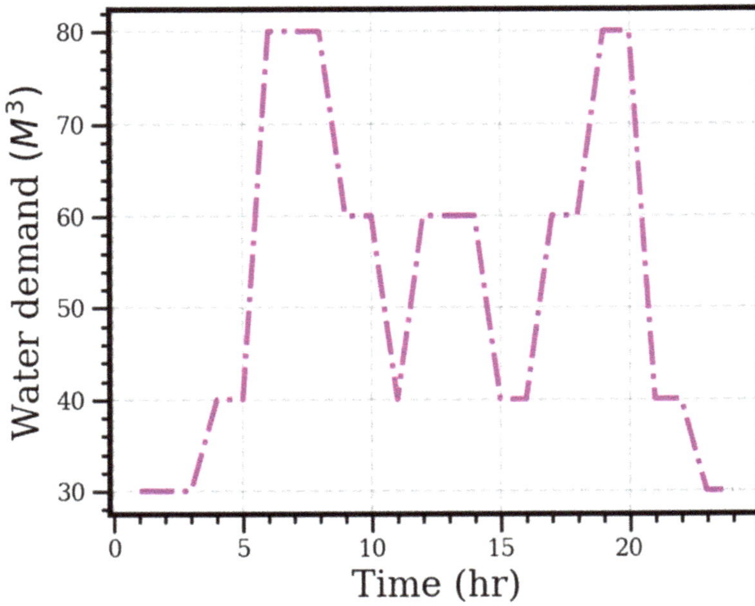

Figure 6.4. Daily water usage pattern [1, 2]. Reprinted from [2], copyright (2021) with permission from Elsevier.

Table 6.1. Results of the daily power supply. Reprinted with permission from [1], copyright (2020) Emerald Publishing Ltd.

Optimised power output (kW)	Case 1: without RES	Case 2: with RES	Case 3: with RES and DR
Grid import power	2670	1748	1322
Grid export power	–	3061	2570
DG output power	1080	1094	1104
PV output power	–	4721	5571

Scenario 1 data shows that the RO unit was mostly powered by the grid, with limited backup from the DG. In this situation, there is no surplus energy to sell back to the grid. This means that even during peak hours, the RO unit is dependent on grid power and the DG to meet load demand, resulting in a high LCOE of 1.56 \$ kWh^{-1}. As shown in figure 6.8 and table 6.2, scenario 1 has the greatest unit cost of freshwater generated with the smallest amount of 1250 m^3.

The results of scenario 2 are better than scenario 1, with imported grid electricity costs falling from \$65 744 in scenario 1 to \$61 172 in scenario 2. This decrease is due to the incorporation of PVs (a renewable energy source) into scenario 2's energy mix. Furthermore, the effect of RES integration is evident in the surplus energy production sold back to the grid, which amounts to a \$16 618 yearly cost.

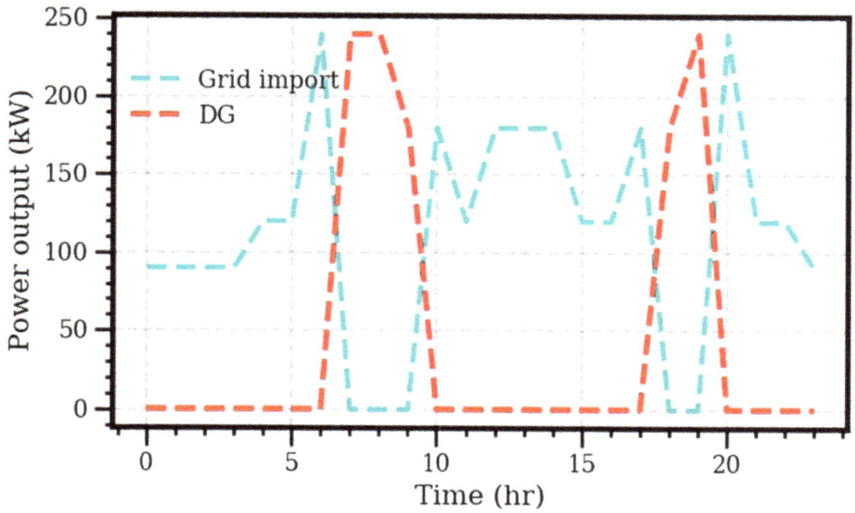

Figure 6.5. Scenario 1: average day energy dispensed. Reprinted with permission from [1], copyright (2020) Emerald Publishing Ltd.

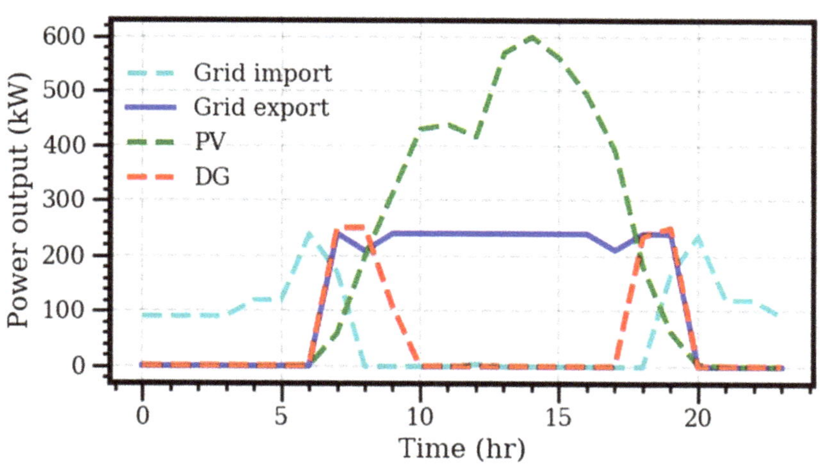

Figure 6.6. Scenario 2: average energy dispensed. Reprinted with permission from [1], copyright (2020) Emerald Publishing Ltd.

Scenario 2 also has a lower LCOE (0.86 $\$ kWh^{-1}$) than case 1 (1.56 $\$ kWh^{-1}$). Finally, with a daily capacity of 1501 m^3/day, scenario 2 generates more freshwater than scenario 1, which has a daily capacity of 1250 m^3/day.

Scenario 3 is a system model that incorporates a ToU DR programme, as well as the integration of three energy sources (grid, DG, and PV) as its energy mix to power the RO unit. Figure 6.7(b) shows the effect of the ToU DR programme by reducing the demand curve. Table 6.2 compares other optimised findings to scenario 1 and 2. Scenario 3 produces more freshwater than scenario 1 and 2 combined, with

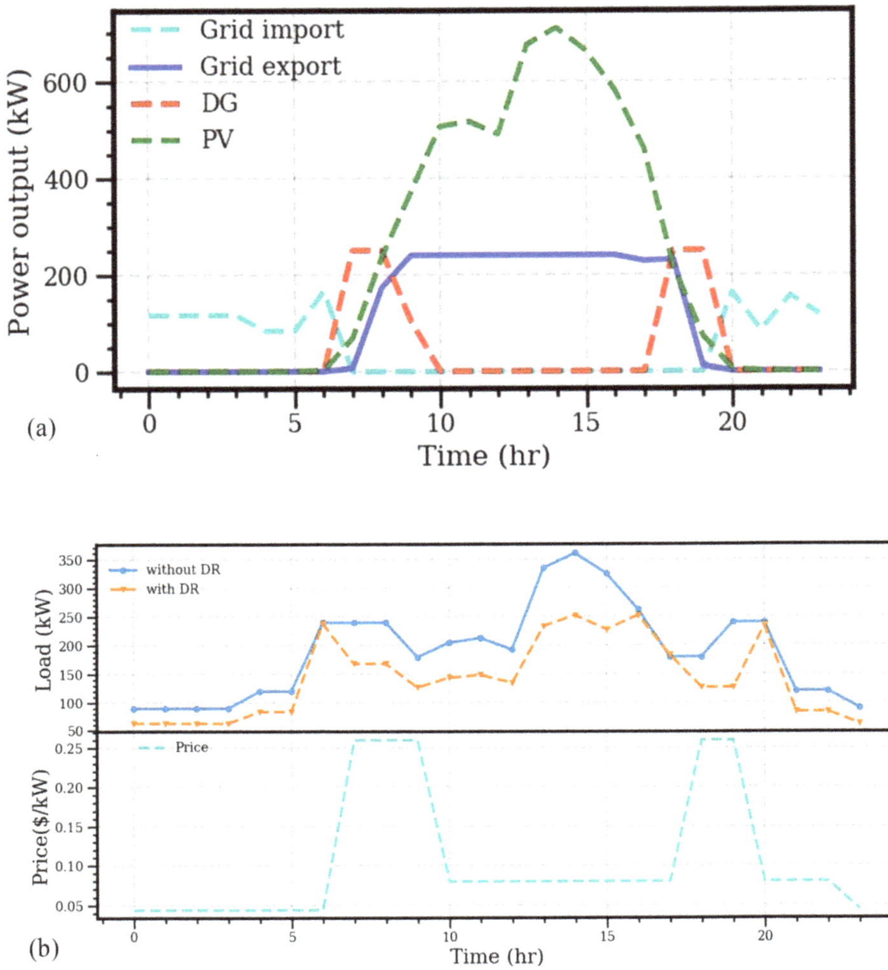

Figure 6.7. (a) Scenario 3: average day energy dispensed. (b) Case 3: hourly load curve with and without DR programme compared to hourly price of power. Reprinted with permission from [1], copyright (2020) Emerald Publishing Ltd.

a daily capacity of 1521 m^3/day and a low cost of 1.36 \$ m^{-3}. Furthermore, scenario 3 has the lowest yearly imported grid power cost of \$26 522 and the highest export grid power cost of \$118 534, lowering the transferable power cost and, as a result, the LCOE to as low as 0.83 \$ kWh^{-1}, compared to 1.56 \$ kW/h^{-1} and 0.86 \$ kWh^{-1} in scenario 1 and 2, respectively. As a result, scenario 3 is the best option due to its largest production capacity at the lowest cost.

6.3 Effect emission cost on water desalination

Most desalination facilities around the world continue to rely heavily on fossil energy sources, particularly the grid. The direct result is large greenhouse gas

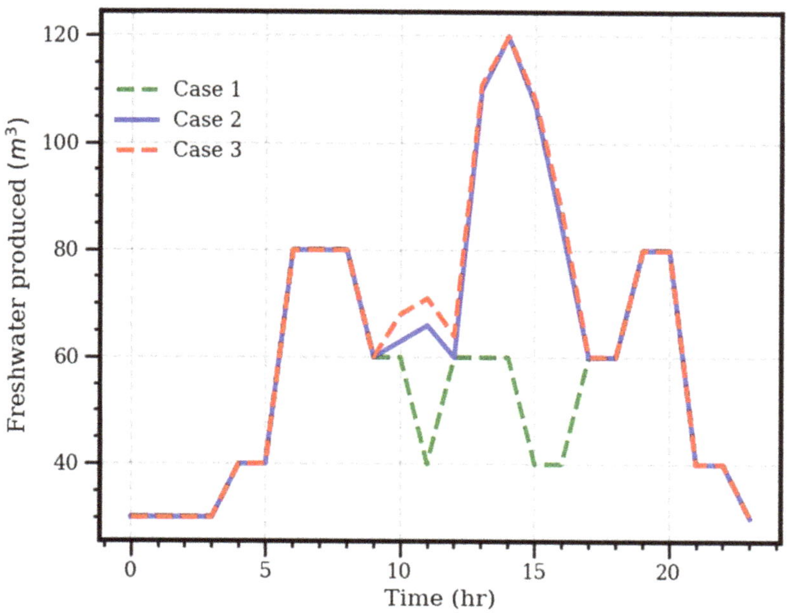

Figure 6.8. Comparison of hourly freshwater production quantity. Reprinted with permission from [1], copyright (2020) Emerald Publishing Ltd.

Table 6.2. Summary of the results of optimised parameters. Reprinted with permission from [1], copyright (2020) Emerald Publishing Ltd.

Optimised parameters	Case 1 without RES	Case 2 with RES	Case 3 with RES and DR
Area of PV		3614	4264
Annual DG fuel cost ($)	275 940	276 594	277 486
Annual grid power imported cost ($)	65 744	61 172	26 522
Annual grid power exported cost ($)		16 618	118 534
ACS ($)	764 802	728 008	752 353
LCOE ($ kWh^{-1})	1.56	0.86	0.83
Daily freshwater produced (m^3/day)	1250	1501	1521
Unit COW ($ m^{-3})	1.68	1.33	1.36

emissions, which have financial and environmental consequences. The cost of exhaled carbon gas is considered one of the external costs of desalination, along with brine [27]. As a result, efficient management of energy supply is critical to reducing gas emissions [28] and, therefore, the cost of desalination.

6.3.1 Case study 2

Case study 2 is an extension of the previous case study to accommodate and account for the cost implications from carbon emissions of desalination systems. Therefore, the three scenarios of the previous case study are repeated to include the cost of carbon emissions and the results are compared.

6.3.2 The economic impact of global warming

The impact of global warming—in other words, the carbon emission of the fossil fuel energy source—could be due to the construction or operation of the plant, with the impact due to the operation exceeding that due to construction by multiple orders [28]. Thus, this study considered the operation as it was adopted by the authors of [28]. The authors defined the impact of global warming as the summation of input power $P_{j,t}$ of every fossil fuel unit j in every time step $t \in T$ multiplied by the specific operational emission factor $\mathrm{SEF}_{j,t}$ and the period Δt_t of the time step, as represented in (6.33)

$$\mathrm{GWI}_g = \sum_{t \in T} \left[\Delta t_t \sum_j \mathrm{SEF}_{j,\,t} \times P_{j,\,t} \right] \tag{6.33}$$

where

$$\mathrm{SEF}_{j,\,t} = \frac{\text{emission due to electrical energy generated}}{\text{electrical energy generated}}. \tag{6.34}$$

The DG emission factor ($\mathrm{SEF}_{\mathrm{DG}}$) depends on its characteristics and fuel consumption, with values ranging from 2.4–2.8 kg l^{-1}; in this study, 2.6 kg l^{-1} is used [10]. Therefore, the DG carbon emission—in other words, the DG GWI—is given as

$$\mathrm{GWI}_{\mathrm{DG}} = \sum_{t=1}^{8760} \mathrm{FC}(t) \times \mathrm{SEF}_{\mathrm{DG}}. \tag{6.35}$$

To estimate the carbon emission cost, the formulated model given by references [27, 29] was adopted, as shown in (6.36)

$$\begin{aligned} \mathrm{CE}\left(\frac{\$}{\mathrm{m}^3}\right) &= \text{Energy supply}\left(\frac{\mathrm{kwh}}{\mathrm{m}^3}\right) \\ &\times \text{Emission factor}\left(\frac{\mathrm{kgCO}_{2-e}}{\mathrm{kwh}}\right) \times \text{Carbon tax}\left(\frac{\$}{\mathrm{kgCO}_{2-e}}\right). \end{aligned} \tag{6.36}$$

For a system with grid and DG units, the total impact of global warming and total carbon emission cost will amount to the sum of the emission from the two units and the sum of the associated cost from the units, as represented in (6.37) and (6.38), respectively

$$\mathrm{GWI} = \mathrm{GWI}_g + \mathrm{GWI}_{\mathrm{DG}} \tag{6.37}$$

$$\mathrm{TEC} = \mathrm{CE}_g + \mathrm{CE}_{\mathrm{DG}}. \tag{6.38}$$

The grid-specific emission factor and carbon tax depend on the country of location of the plant. This study adopted the calculated emission factor for South Africa given by reference [30]. The value of the emission factor is 1.069 $kgCO_2\ kwh^{-1}$. Recently, the South African carbon tax rate ranged from R6–R48 ($0.40–$3.17)[31]. In this study, the lowest value of $0.40 is considered.

6.3.3 Optimisation model

Similarly to case study 1, the ACS matrix is used for the economic evaluation of the RO desalination system powered by the different energy sources to determine the cost of freshwater (equation (6.25)) and the LCOE (equation (6.26)). The ACS involves equations (21), (22), and (23). Also included in the ACS are DR cost (ΔLC) and total cost of carbon emissions (TEC)

$$\begin{aligned} \mathrm{ACS} = {} & \mathrm{TICC} \times \mathrm{CRF} + \mathrm{TMC} + \mathrm{AC_{MR}} \times \mathrm{CRF} \\ & + \mathrm{AC_{CH}} + \mathrm{CGP} + \mathrm{FC_{DG}} + \Delta\mathrm{LC} + \mathrm{TEC}. \end{aligned} \tag{6.39}$$

The optimisation problem is formulated as a multi-objective linear programming problem as expressed in (6.40), minimising ACS and carbon emission while maximising the quantity of freshwater production subject to the same constraints expressed in (6.28)–(6.32). Weighting factors ($w_1 = 0.3$, $w_2 = 0.5$, and $w_3 = 0.2$) were assigned based on the preference of major concern, with the GWI of emission ranking highest, then the ACS and quantity of water produced ranking lowest, since meeting water demand is more important than excess water production. This multi-optimisation problem is solved using the CPLEX solver of AIMMS.

Objective Function

$$\mathrm{Min}\left[w_1 \times \mathrm{ACS} + w_2 \times \sum \mathrm{GWI} - w_3 \times \sum \mathrm{QW}(t)\right] \tag{6.40}$$

S.t. The same constraint as (6.28–6.32) in the previous case study. The flow chart is similar to that in figure 6.1.

6.4 Results and discussion

Figure 6.9 shows the hourly power supply from only the grid, while figure 6.10 represents the hourly power output of the energy supply mix of scenario 2 and the excess power exported to the grid. Figure 6.11 represents the hourly power supply mix for scenario 3, while figure 6.12(b) depicts the difference between the load curve before and after the application of the DR programme. The volume of hourly water produced by each case is as shown in figure 6.13, while table 6.3 is a summary of the results.

Figure 6.9 shows the influence of ToU DRs on moving demand from peak to off-peak hours, hence lowering the cost of imported electricity from the grid. As stated in table 6.4, the cost of imported grid power is lowered from $168 236 to $161 140. Scenario 1 has a lower ACS, LCOE, and COW production when compared to scenario 2, but the high carbon emissions of 146 562 $kgCO_{2\text{-}e}$ and the related cost of $600 471, as shown in figure 6.14, are disappointing.

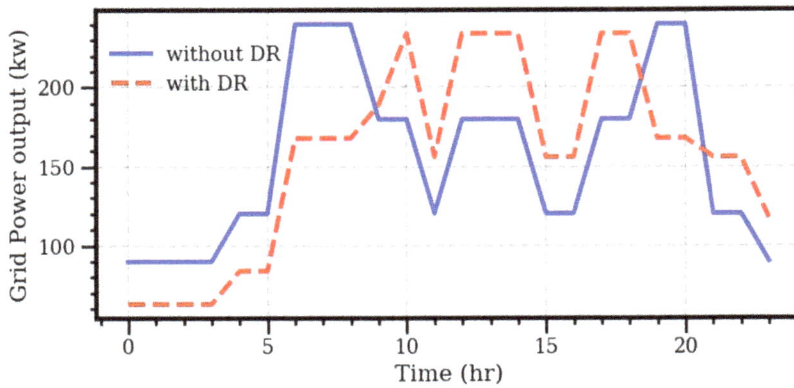

Figure 6.9. Scenario 1: hourly energy dispensed from the grid [2]. Reprinted from [2], copyright (2021), with permission from Elsevier with permission from Elsevier.

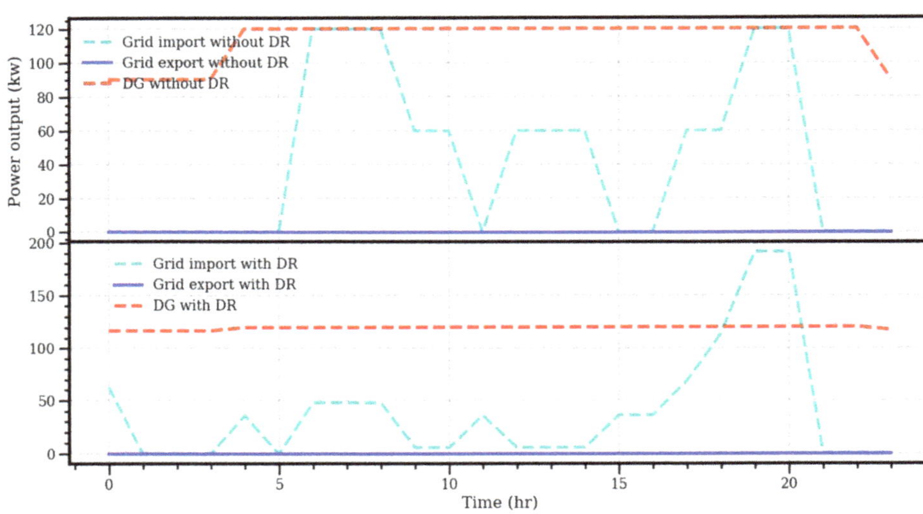

Figure 6.10. Scenario 2: hourly energy dispensed. Reprinted from [2], copyright (2021), with permission from Elsevier.

In figure 6.10 and table 6.3, the results of scenario 2 demonstrate a considerable decrease in imported electricity when compared to scenario 1. This decrease is due to the electricity distribution between the grid and DG. The ToU DR programme, as illustrated in figure 6.11, achieves a significant decrease. As a result, DR optimally reschedules the power supply from the grid and DG to reduce carbon emissions and costs. Scenario 2, with the highest cost of ACS, LCOE, and water production cost, has a lower carbon emission and emission cost of 1234 010 $kgCO_{2\text{-}e}$ and \$505 944, respectively, than scenario 1. DR reduces these figures to 1181 286 $kgCO_{2\text{-}e}$ and \$493 019, as shown in figure 6.14 and tables 6.5 and 6.4.

Scenario 3, an RO desalination system powered by the grid, DG, and RES (PV), exhibits a considerable improvement in the reduction of carbon emissions and

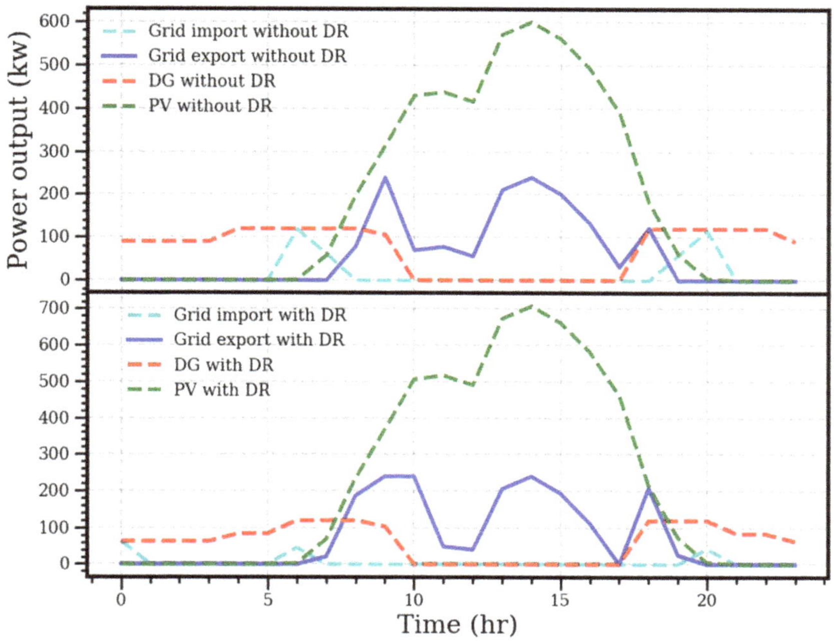

Figure 6.11. Scenario 3: hourly energy dispensed. Reprinted from [2], copyright (2021), with permission from Elsevier.

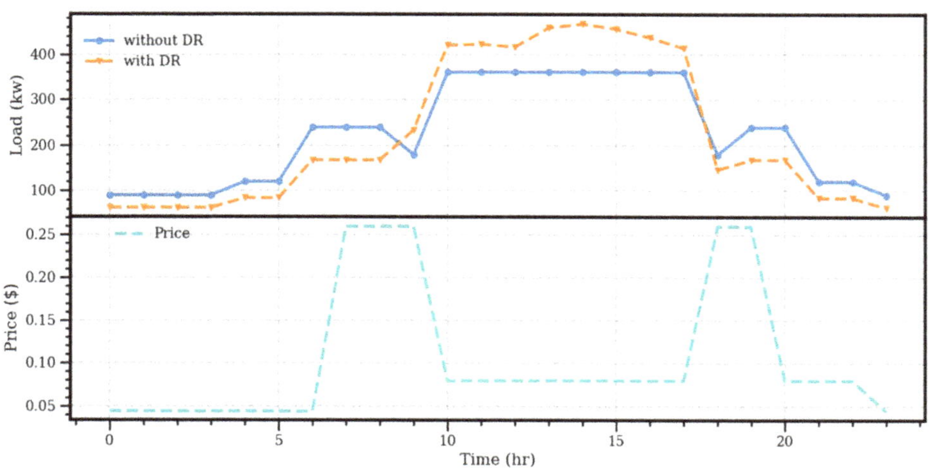

Figure 6.12. Scenario 3: hourly load curve with and without DR programme compared with hourly price of power. Reprinted from [2], copyright (2021), with permission from Elsevier.

overall system cost. Scenario 3's power output, as shown in figure 6.11, indicates a significant drop in grid-imported power to just two hours per day, with a total output of 354 kW and a further reduction by DR to 89.2 kW. Furthermore, when scenario 3 is compared to scenario 2, the output power from DG is reduced. These

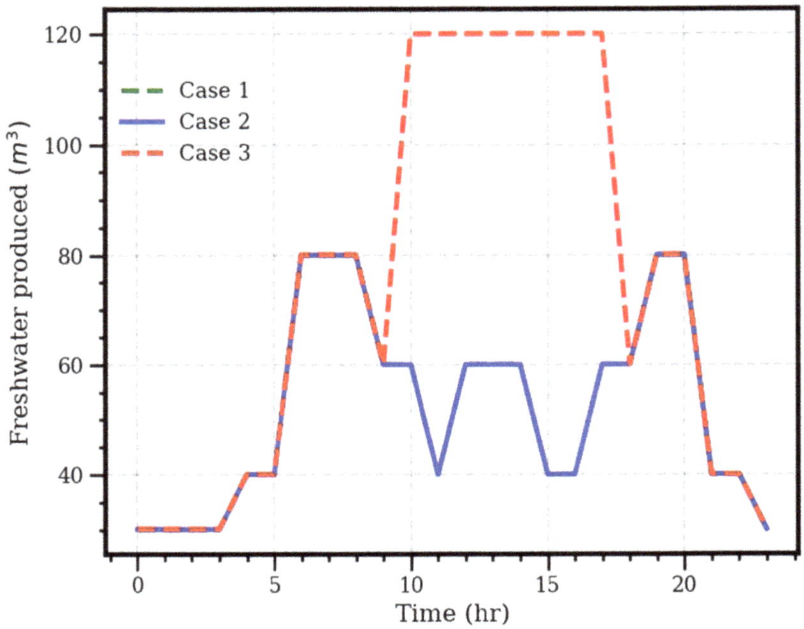

Figure 6.13. Comparison of hourly freshwater production quantity. Reprinted from [2], copyright (2023), with permission from Elsevier.

Table 6.3. Summary of power outputs. Reprinted from [2], copyright (2021), with permission from Elsevier with permission from Elsevier.

Optimised power output (kWh)	Case 1: grid only		Case 2: grid and DG		Case 3: grid, DG, and PV	
	Without DR	With DR	Without DR	With DR	Without DR	With DR
Grid import power	3750	3750	1020	885	354	89.2
Grid export power	–	–	–	–	1461	1765
DG output power	–	–	2730	2865	1756	1475
PV output power	–	–	–	–	4721	5571

decreases in imported grid power and DG power output are the consequence of the integration of an RES (PV), which has a high power output at particular times of the day. This has resulted in an excess of overall power generation, which is sold back to the grid.

In addition to the decrease in grid and DG power supply as a consequence of the integration of an RES and ToU DR in scenario 3, the effect of the DR programme is seen in the reduction of carbon emissions and costs, as shown in figure 6.12. The

Table 6.4. Summary results of cost parameters. Reprinted from [2], copyright (2021), with permission from Elsevier with permission from Elsevier.

Optimised cost ($)	Case 1: grid only		Case 2: grid and DG		Case 3: grid, DG, and PV	
	Without DR	With DR	Without DR	With DR	Without DR	With DR
Grid import cost	168 236	161 140	59 743	51 544	16 652	1993
Grid export cost	–	–	–	–	71 677	96 335
DG fuel cost	–	–	334 260	346 578	245 398	219 712
Grid emission cost	600 471	600 471	163 328	141 711	56 691	14 276
DG emission cost	–	–	342 616	342 616	251 533	251 533
Total emission cost	600 471	600 471	505 944	493 019	308 224	265 809

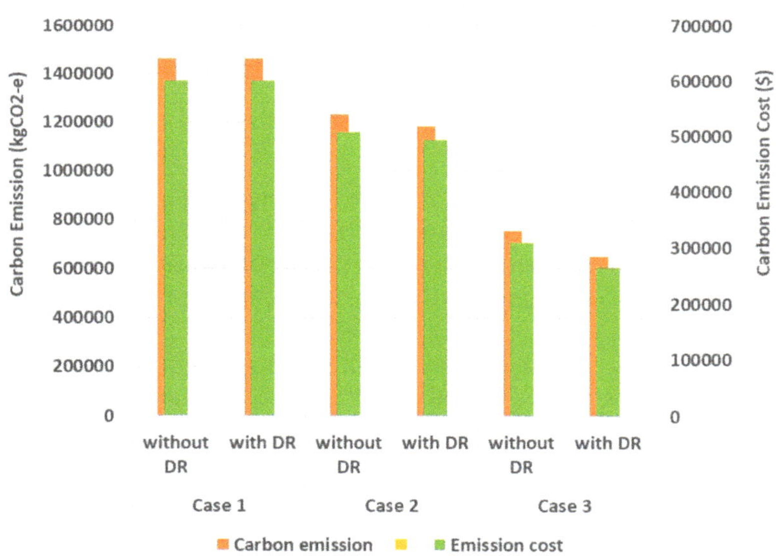

Figure 6.14. Comparison of emissions produced and the associated cost for the three cases. Reprinted from [2], copyright (2021), with permission from Elsevier.

demand curve in figure 6.13 also demonstrates a change in load from peak to off-peak hours, and the resulting optimised outcome compared to scenarios 1 and 2 is presented in tables 6.4 and 6.5. Scenario 3 produces more freshwater than scenarios 1 and 2, with an average daily volume of 1790 m^3 at a low unit cost of 1.92 $ m^{-3} without DR and 1.77 $ m^{-3} with DR. Furthermore, as shown in table 6.4, scenario 3

Table 6.5. Summary of optimised results. Reprinted from [2], copyright (2021), with permission from Elsevier with permission from Elsevier.

Optimised parameters	Case 1: grid only		Case 2: grid and DG		Case 3: grid, DG, and PV	
	Without DR	With DR	Without DR	With DR	Without DR	With DR
ACS ($)	1203 524	1203 524	1337 972	1326 506	1255 694	1158 801
Emission ($kgCO_{2\text{-}e}$)	1464 563	1464 562	1234 010	1181 286	751 766	648 315
Daily water produced (m^3)	1250	1250	1250	1250	1790	1790
LCOE ($ kWh^{-1})	1.88	1.88	1.98	1.97	1.29	1.20
COW ($ m^{-3})	2.64	2.64	2.93	2.91	1.92	1.77

has the lowest annual DG fuel cost, the lowest annual grid power imported, and is the only instance with exported grid power cost, resulting in the lowest ACS. As a result, scenario 3's LCOE and COW produced are lower than those of scenarios 1 and 2; thus, scenario 3 generates more freshwater at a lower cost than scenarios 1 and 2, as shown in table 6.4.

6.4.1 Sensitivity analysis

The three examples with and without DR are subjected to a sensitivity analysis. The findings in figures 6.15 and 6.16 indicate the impact of a higher carbon tax rate on LCOE and the cost of the generated water. The carbon tax rate and the LCOE have a direct proportional connection. Furthermore, the cost of producing water tends to rise in tandem with an increase in the price of carbon.

Table 6.6 highlights the influence of DR in the model once again, demonstrating that the increase in both LCOE and COW as a consequence of a carbon price increase is much greater in situations without DR compared to cases with DR.

6.5 Summary

In this chapter, two mathematical optimisation models are developed to compare a standard grid and DG with a system that incorporates PVs, a renewable energy source, to power an RO desalination unit. The first case study model is designed with a balance of electricity supply and demand, water demand and production, cost functions, and the ToU DR programme in mind. Three situations were used to test the model. In scenario 1, the grid and a DG are powered by an RO desalination system if conventional energy sources are used. The model for this scenario does not include any RES in the energy mix and does not use a DR mechanism. The second scenario included a PV and a renewable energy source, with conventional energy sources (grid and DG), but did not include a DR programme. Finally, scenario 3 included a grid-powered RO desalination plant, a DG, PVs, and a ToU DR

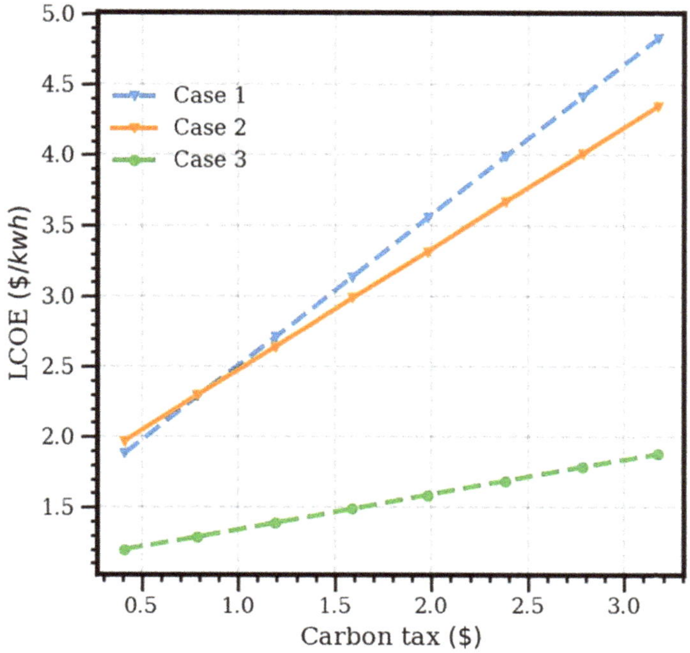

Figure 6.15. Effect of the increase in the carbon tax on the LCOE for the three cases with DR. Reprinted from [2], copyright (2021), with permission from Elsevier.

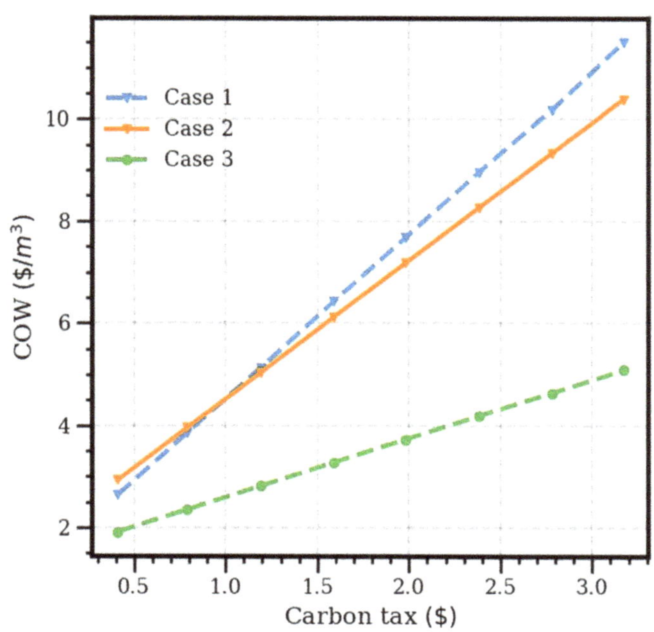

Figure 6.16. Effect of the increase in the carbon tax on the COW for the three cases with DR. Reprinted from [2], copyright (2021), with permission from Elsevier with permission from Elsevier.

Table 6.6. Result of cost parameters. Reprinted from [2], copyright (2021), with permission from Elsevier with permission from Elsevier.

	Case 1				Case 2				Case 3			
	Without DR		With DR		Without DR		With DR		Without DR		With DR	
C. tax	LCOE	COW	LCOE	COW	LCOE	COW	LCOE	COW	LCOE	COW	LCOE	COW
0.4	1.88	2.64	1.88	2.64	1.98	2.93	1.97	2.93	1.29	1.92	1.20	1.77
0.79	2.29	3.86	2.29	3.86	2.32	3.96	2.30	3.89	1.40	2.36	1.29	2.15
1.19	2.71	5.14	2.71	5.14	2.68	5.04	2.64	4.93	1.52	2.82	1.39	2.55
1.59	3.14	6.43	3.14	6.43	3.04	6.12	2.99	5.96	1.65	3.28	1.49	2.94
1.98	3.56	7.68	3.56	7.68	3.39	7.18	3.32	6.97	1.76	3.73	1.59	3.33
2.38	3.99	8.96	3.99	8.96	3.75	8.26	3.67	8.01	1.88	4.19	1.69	3.73
2.78	4.42	10.2	4.42	10.2	4.11	9.34	4.01	9.04	2.00	4.65	1.79	4.13
3.17	4.83	11.5	4.83	11.5	4.47	10.4	4.35	10.05	2.12	5.10	1.88	4.51

programme on the demand side. These models aim to optimise freshwater production at the lowest possible cost and estimate the influence of DR and its cost on the amount of water produced at various times of the day and the unit cost of freshwater. The findings reveal that scenario 3 greatly increases freshwater production at a much lower cost than the other two examples. As a result, scenario 3 is suggested as the most appropriate size option for the RO desalination system. The freshwater cost and LCOE are also lower than those of comparable situations in the literature, such as 2.962 and 6.457 \$ m^{-3} from [32], 1.08–1.5076 \$ m^{-3} in [10], and 1.9452–2.0196 \$ m^{-3} in [11].

Case study 2 reproduced the three scenarios from case study 1, including the carbon emissions cost as one of the external expenses of desalination. The three scenarios were techno-economically analysed using the LCOE, COW, and ACS cost matrices, and the findings were compared. The findings revealed that scenario 3 provided a very substantial improvement in terms of carbon emission reduction and overall system cost. Therefore, scenario 3 is proposed as the optimal optimisation model for the RO desalination system, demonstrating that more freshwater can be generated at the lowest possible cost. This proposal validates RES as viable alternatives to fossil fuel power sources. On the other hand, the incorporation of the DR programme improves the ideal schedule of energy supply schedule to meet the demand of the desalination unit, while reducing cost and carbon emissions and boosting freshwater production.

References

[1] Okampo E J and Nwulu N 2020 Optimal energy mix for a reverse osmosis desalination unit considering demand response *J. Eng. Des. Technol.* **18** 1287–303

[2] Okampo E J and Nwulu N 2021 Techno-economic evaluation of reverse osmosis desalination system considering emission cost and demand response *Sustain. Energy Technol. Assess.* **46** 101252

[3] Al-Karaghouli A and Kazmerski L L 2013 Energy consumption and water production cost of conventional and renewable energy-powered desalination processes *Renew. Sustain. Energy Rev.* **24** 343–56

[4] Mezher T, Fath H, Abbas Z and Khaled A 2011 Techno-economic assessment and environmental impacts of desalination technologies *Desalination* **266** 263–73

[5] Maleki A, Pourfayaz F and Ahmadi M H 2016 Design of a cost-effective wind/photovoltaic/hydrogen energy system for supplying a desalination unit by a heuristic approach *Sol. Energy* **139** 666–75

[6] Spyrou I D and Anagnostopoulos J S 2010 Design study of a stand-alone desalination system powered by renewable energy sources and a pumped storage unit *Desalination* **257** 137–49

[7] Okampo E J and Nwulu N 2021 Optimisation of renewable energy powered reverse osmosis desalination systems: a state-of-the-art review *Renew. Sustain. Energy Rev.* **140** 110712

[8] Okampo E J and Nwulu N 2022 Economic and reliability assessment of hybrid PRO-RO desalination systems using brine for salinity gradient energy production *Sustainability* **14** 3328

[9] Ghenai C, Merabet A, Salameh T and Pigem E C 2018 Grid-tied and stand-alone hybrid solar power system for desalination plant *Desalination* **435** 172–80

[10] Abdelshafy A M, Hassan H and Jurasz J 2018 Optimal design of a grid-connected desalination plant powered by renewable energy resources using a hybrid PSO-GWO approach *Energy Convers. Manag.* **173** 331–47
[11] Wu B, Maleki A, Pourfayaz F and Rosen M A 2018 Optimal design of stand-alone reverse osmosis desalination driven by a photovoltaic and diesel generator hybrid system *Sol. Energy* **163** 91–103
[12] Nwulu N I and Xia X 2017 Optimal dispatch for a microgrid incorporating renewables and demand response *Renew. Energy* **101** 16–28
[13] Okampo E J and Nwulu N 2020 Optimal design and techno-economic evaluation of renewable energy powered combined reverse osmosis desalination and brine treatment unit *Desalin. Water Treat.* **202** 27–37
[14] Peng W, Maleki A, Rosen M A and Azarikhah P 2018 Optimization of a hybrid system for solar-wind-based water desalination by reverse osmosis: comparison of approaches *Desalination* **442** 16–31
[15] Smaoui M, Abdelkafi A and Krichen L 2015 Optimal sizing of stand-alone photovoltaic/wind/hydrogen hybrid system supplying a desalination unit *Sol. Energy* **120** 263–76
[16] Maleki A, Khajeh M G and Rosen M A 2016 Weather forecasting for optimisation of a hybrid solar-wind–powered reverse osmosis water desalination system using a novel optimiser approach *Energy* **114** 1120–34
[17] Majidi M, Mohammadi-Ivatloo B and Anvari-Moghaddam A 2019 Optimal robust operation of combined heat and power systems with demand response programmes *Appl. Therm. Eng.* **149** 1359–69
[18] Jabari F, Jabari H, Mohammadi-ivatloo B and Ghafouri J 2019 Optimal short-term coordination of water-heat-power nexus incorporating plug-in electric vehicles and real-time demand response programmes *Energy* **174** 708–23
[19] Dehnavi E and Abdi H 2016 Optimal pricing in time of use demand response by integrating with dynamic economic dispatch problem *Energy* **109** 1086–94
[20] Acuña L G, Lake M, Padilla R V, Lim Y Y, Ponzón E G and Soo Too Y C 2018 Modelling autonomous hybrid photovoltaic wind energy systems under a new reliability approach *Energy Convers. Manag.* **172** 357–69
[21] Al-Obaidi M A, Filippini G, Manenti F and Mujtaba I M 2019 Cost evaluation and optimisation of hybrid multi-effect distillation and reverse osmosis system for seawater desalination *Desalination* **456** 136–49
[22] Koroneos C, Dompros A and Roumbas G 2007 Renewable energy driven desalination systems modelling *J. Clean. Prod.* **15** 449–64
[23] Eskom 2019 *Schedule of Standard Prices for Eskom Tariffs 1 April 2019 to 31 March 2020 for Non-Local Authority Supplies, and 1 July 2019 to 30 June 2020 for Local Authority Supplies*
[24] Du Plessis J A, Burger A J, Swartz C D and Musee N 2006 A desalination guide for South African municipal engineers *Water SA* **32** 641–7
[25] Goga T, Friedrich E and Buckley C A 2015 *A LCA (Life Cycle Assessment) Comparison of Wastewater Reclamation and Desalination for the Ethekwini Municipality—A Theoretical Study*
[26] Department of Water and Sanitation, City of Cape Town 2018 *Water Outlook 2018 Report*
[27] Molinos-Senante M and González D 2018 Evaluation of the economics of desalination by integrating greenhouse gas emission costs: an empirical application for Chile *Renew. Energy* **133** 1327–37

[28] Baumgärtner N, Delorme R, Hennen M and Bardow A 2019 Design of low-carbon utility systems: exploiting time-dependent grid emissions for climate-friendly demand-side management *Appl. Energy* **247** 755–65
[29] Kesieme U K, Milne N, Aral H, Yong C and Duke M 2013 Economic analysis of desalination technologies in the context of carbon pricing, and opportunities for membrane distillation *DES* **323** 66–74
[30] Brander M, Sood A, Wylie C, Haughton A and Lovell J 2011 *Electricity-Specific Emission Factors for Grid Electricity*
[31] Media Statement: Publication of the 2019 Carbon Tax ActNational Treasury, Republic of South Africa 2019
[32] Gökçek M and Gökçek Ö B 2016 Technical and economic evaluation of freshwater production from a wind-powered small-scale seawater reverse osmosis system (WP-SWRO) *Desalination* **381** 47–57

Chapter 7

Optimal design and evaluation of combined desalination and brine treatment units

The rapid proliferation of desalination plants to address water scarcity challenges has brought to light a significant environmental concern—the disposal of concentrated brine by-product. Desalination, while providing a solution to freshwater scarcity, generates brine as a concentrated waste stream containing elevated levels of salt and other contaminants. The disposal of brine into oceans, seas, or inland water bodies can disrupt local ecosystems, leading to potential consequences for marine life, water quality, and coastal habitats. This chapter explores emerging solutions and mitigation strategies aimed at minimising the environmental footprint of brine disposal. It highlights the potential of brine concentration technologies, beneficial reuse options, and hybrid approaches to reduce the volume and impact of discharged brine, offering a more sustainable pathway for managing this waste stream. To achieve this, the chapter presents two case studies that utilise hybrid desalination technologies for both water treatment and brine utilisation. While case study 1 models a system that combines reverse osmosis (RO), electro-dialysis (ED), and crystallisation (CRY) methods of desalination for freshwater and salt production, case study 2 combines RO and pressure-retarded osmosis techniques for both water treatment and energy generation. The result from both systems minimises brine production and by extension reduces the COW production. These integrated systems of desalination enhance environmental sustainability by efficient utilisation of brine rather than the hazardous effect of its disposal.

7.1 Introduction

Seawater often has a high salt content, which makes it unpleasant and harmful to consume. Desalination is the process of obtaining freshwater from seawater. This process leaves behind more concentrated salty water known as brine. Ecology is

doi:10.1088/978-0-7503-6154-5ch7

harmed by indiscriminate brine disposal. This harmful effect increases the external costs of desalination alongside the impact of global warming [1]. As a result, the primary obstacles to seawater desalination are energy consumption and brine control. These obstacles make saltwater desalination prohibitive in terms of cost and the environmental effect of the energy source and the disposal of the brine. Therefore, effective energy supply and brine management are critical to reducing carbon emissions and evaluating the cost of desalination [2].

In the previous chapter, we developed an optimisation model to evaluate and minimise the problem of carbon emissions from a desalination system. This was achieved by using renewable energy sources to complement traditional energy sources [3–5]. In this chapter, we attempt to further extend the cost evaluation of a desalination system to include brine management.

One major limitation of RO desalination technology is the volume of brine (concentrate) produced during the desalination process [6]. Different brine management approaches have been proposed and some have been implemented [7, 8]. Some of the brine treatment technologies, such as desalination technologies, are thermal or membrane-based. Most often, the same techniques are used in a specific combination for both desalination and brine treatment.

In this chapter, case study 1 presents a hybrid desalination technique which comprises RO, ED, and CRY in an integrated desalination and brine treatment unit. This integrated system is designed so that RO is employed for desalination, whereas ED and CRY are utilised for further concentrate treatment. Because ED is a membrane technique, it is inexpensive for brine treatment since it employs ion selectivity for separating freshwater from the concentrate. This method does not completely change brine to drinkable water and salt. Scaling of soluble salt on the membrane limits its efficacy. As a result, further treatment of the highly concentrated brine is necessary. The CRY of the remaining brine volume may result in zero brine discharge since the final product is salt crystals and evaporated freshwater, both of which can be condensed and collected for drinking. Reference [9] showed the viability of RO-ED-CRY, and a model of framework was built for the cost and energy requirements. A comparison was made between RO-ED-CRY and the same system with a two-stage high-pressure RO; the findings reveal that adding high-pressure RO is cost prohibitive.

Furthermore, case study 2 of this chapter presents a hybrid of high-pressure RO and RO for proper brine management from seawater desalination, following a similar approach to forward osmosis (FO)-RO integration in [10–12]. While [10, 11] focus on the thermodynamic process optimisation of the hybrid systems, [12] extends this to the economic assessment of the system using the life cycle cost index. This research considers the reduction of carbon emissions and their related costs using levelized cost of energy (LCOE), annualised cost of the system (ACS), and cost of water (COW) as economic indicators. The loss-of-load probability (LOLP) index is used in this research to measure the dependability of the pressure-retarded osmosis (PRO)-RO hybrid system while optimising energy scheduling.

7.1.1 Environmental impacts and management of brine

When desalinating freshwater, one of the primary concerns is how brine will be treated and what its impacts will be on the environment. Brine is a by-product of desalination. It has an extremely high percentage of salinity and can include a variety of compounds, including $FeCl_3$, H_2SO_4, and more [8]. It is considered an additional cost of desalination to manage brine in a way that prevents it from having a detrimental influence on the surrounding environment [1]. At this time, a wide range of desalination plants discharge brine output back into the bodies of water closest to the location where the plant was installed. This technique has very negative effects on ocean wildlife [13, 14]. The brine may also be disposed of by spreading it out on the ground surface or digging trenches. Because of the high temperature of the brine, these approaches are not considered feasible options. This is because brine has the potential to damage arable land that may be utilised for other reasons. Procedures also have a high financial investment. Several researchers are in agreement about the severity of the environmental damage that careless dumping of brine can cause and have further offered a variety of treatment approaches to lessen the severity of the problem [15–17]. Roberts *et al* [15] conducted an in-depth analysis of the effects that brine discharges have on the marine environment. Based on their findings, they recommended the establishment of a control-impact testing and monitoring system to determine the extent of the impact before and after the desalination process. When completing an evaluation of the desalination process, Lior [16] suggested adopting a comprehensive approach to a paradigm sustainability model that considered the effects on social and economic systems and the environment. It has been proposed that a sustainable strategy for brine management would include pretreatment of feed water, the treatment of brine after it has been produced, and the use of hybrid desalination techniques [10–12]. This approach has the added benefit of energy generation and efficient utilisation.

When considering seawater desalination for sustainability and environmental friendliness, the cost and correct management of the energy supply and brine production must be considered [18]. The subsequent case studies in this chapter present optimisation models that combine desalination of freshwater, brine treatment, and generation of renewable energy with the objectives of cost minimisation and the reduction of the effects of carbon emissions and brine disposal.

7.2 Case study 1: hybridisation of RO-ED-CRY

7.2.1 System architecture

The architecture of the system can be seen shown in figure 7.1. The system is divided into two primary parts: (i) the power section, which combines renewable energy sources such as wind and photovoltaics (PVs) with grid power; and (ii) the desalination and brine treatment section, which combines RO, ED, and CRY units. The power sources are optimally scheduled with time-of-use (ToU) demand response (DR) systems and measure carbon emissions in order to achieve the lowest possible LCOE. Meanwhile, the desalination and brine treatment section is designed

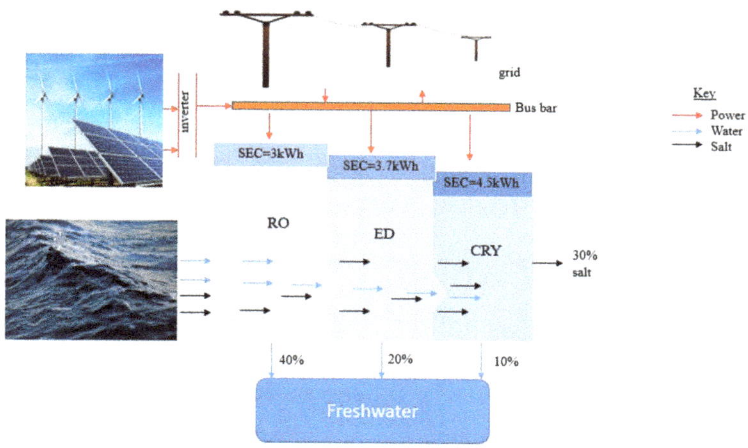

Figure 7.1. Schematic diagram of the combined desalination and brine treatment process. Reprinted with permission from [18].

in such a way that the RO unit desalinates 40% of the feed water (seawater) into freshwater, leaving 60% as brine according to equations (6.5)–(6.7), which was adopted from [8, 19]. This brine is then sent through ED, which converts an additional 20% of the feed water into freshwater (i.e. 33.3333% of the brine that was sent through ED), resulting in a brine that is more concentrated once it is returned to the CRY unit. The CRY unit is responsible for evaporating 16.6667% of the initial RO brine, resulting in an additional 10% of the input water becoming freshwater after condensing. The remaining 30% of the feed water, which accounts for 50% of the RO's initial brine, crystallises into salt, resulting in zero discharge.

7.2.2 Energy source model

The energy sources used to power this integrated desalination and brine treatment system are electricity from the traditional power grid and renewable energy sources of PVs and wind. Implementing the same grid and PV power output models from chapter 6 (equations (6.1)–(6.6)), we shall only detail the wind power output in this chapter to complete the energy source model to be sized and evaluated.

The hourly wind speed ($V(t)$), the swept area of the wind turbine (AWT), and the air density (ρ_{air}) are the necessary parameters to model the wind power output. Other parameters include the efficiency of the wind generator ($\acute{\eta}_w$) and the power coefficient (C_p). Therefore, the hourly power output of the wind generator ($W_p(t)$) is given as [20]

$$W_p(t) = \frac{1}{2} \times \acute{\eta}_w \times \rho_{\text{air}} \times C_p \times \text{AWT} \times V(t)^3 \tag{7.1}$$

$$V(t) = V_R \times \left[\frac{h}{h_R}\right]^{\propto} \tag{7.2}$$

where $V(t)$ is the hourly speed at projected height (h), V_R is the hourly speed at reference height (h_R), and α is the power law exponent equivalent to 1/7 [21].

The cost of using wind power for desalination is similar to that of PVs and, in this study, these are analysed based on the initial capital cost (IC_{WT}) and total maintenance cost (TMC_{WT}), depending on the AWT, as follows:

$$IC_{WT} = AWT \times C_{WT} \tag{7.3}$$

$$AMC_{WT} = AWT \times MC_{WT} \times r. \tag{7.4}$$

where AMC is annual maintenance cost and MC is maintenance cost.

7.2.3 Desalination and brine management energy demand model

The energy demand for RO desalination modelled in chapter 6 (equations (6.13)–(6.17)) is adopted in this chapter as well. The brine treatment section includes the ED and CRY units. The volumetric quantity of brine (QB) produced from the RO plant depends on its water production capacity and water recovery ratio (RR), expressed as [19]

$$DQB = \frac{DQW}{RR} \times (1 - RR) \tag{7.5}$$

where RR is the percentage of the volume of freshwater produced by the RO desalination to the volumetric quantity of saline feed water (QF) [8]; it is assumed to be 40% in this study

$$RR = \frac{QW}{QF} \times 100. \tag{7.6}$$

This implies that QB can also be calculated as [8]

$$QB = QF - QW. \tag{7.7}$$

Equation (7.8) is modified to consider the hourly brine production as follows:

$$QB(t) = \frac{QW_{RO}(t)}{RR} \times (1 - RR). \tag{7.8}$$

Therefore, the hourly power demand ($P_B(t)$) of the brine treatment unit depends on the sum of the specific energy consumption (SEC) by ED and the CRY unit, which in this study are adapted from [9, 22] as 3.7 and 4.5 kwh m^{-3}, respectively. Hence,

$$P_B(t) = QB(t) \times SEC. \tag{7.9}$$

To achieve a zero brine discharge, the brine that is produced by the RO unit must first be processed by the ED unit. This process results in the production of certain amounts of freshwater (in this research, 20% of the brine) as well as a more concentrated brine, which is then processed by the crystalliser. This also resulted in the production of certain quantities of freshwater, which in this instance made up 10% of the total brine generated by the RO unit. The remaining quantities of brine consisted of crystallised salt and other chemicals. The following equation may be used to calculate the total amount of freshwater generated by the combined RO-ED-CRY system:

$$\mathrm{TQW}(t) = \mathrm{QW_{RO}}(t) + \mathrm{QW_{ED}}(t) + \mathrm{QW_{CRY}}(t). \quad (7.10)$$

where TQW is total quantity of water produced. The cost-of-use DR cost is expressed as [23]

$$\Delta \mathrm{LC} = \sum \left[(P_{\mathrm{WD}}(t) + P_B(t)) \times \lambda(t) - \left((P_{\mathrm{WD}}(t) + P_B(t)) - \Delta L(t) \right) \times \lambda(t) \right] \quad (7.11)$$

$$t = 1, 2, 3, \ldots 8760.$$

To account for the global warming impact of the traditional energy sources used to power this integrated system, equations (6.33), (6.34), and (6.36) of chapter 6 are adopted

$$\mathrm{GWI}_g = \sum_{t \in T} \left[\Delta t_t \sum_j \mathrm{SEF}_{j,t} \times P_{j,t} \right]$$

$$\mathrm{SEF}_{j,t} = \frac{\text{emission due to electrical energy generated}}{\text{electrical energy generated}}$$

$$\mathrm{CE}\left(\frac{\$}{\mathrm{m}^3}\right) = \text{Energy supply}\left(\frac{\mathrm{kwh}}{\mathrm{m}^3}\right) \times \text{Emission factor}\left(\frac{\mathrm{kgCO_{2-e}}}{\mathrm{kwh}}\right) \times \text{Carbon tax}\left(\frac{\$}{\mathrm{kgCO_{2-e}}}\right).$$

7.2.4 Optimisation problem formulation

The ACS matrix is used for the economic evaluation of the RO desalination system powered by the different energy sources to determine the cost of freshwater and salt (COP). ACS involves the capital recovery factor (CRF) and the total system cost, which includes total initial capital cost (TICC) and TMC of all the components that make up the system [24–26]. Also included in the ACS are DR cost (ΔLC) and total carbon emission cost (EC).

$$\mathrm{TICC} = \mathrm{IC_{PV}} + \mathrm{IC_{WT}} + \mathrm{IC_{RO}} + \mathrm{IC_{ED}} + \mathrm{IC_{CRY}} + \mathrm{ICW_{TK}} \quad (7.12)$$

$$\mathrm{TMC} = \mathrm{AMC_{PV}} + \mathrm{AMC_{WT}} + \mathrm{AMC_{RO}} + \mathrm{AMC_{ED}} + \mathrm{AMC_{CRY}} \quad (7.13)$$

$$\mathrm{ACS} = \mathrm{TICC} \times \mathrm{CRF} + \mathrm{TMC} + \mathrm{AC_{MR}} \times \mathrm{CRF} + \mathrm{AC_{CH}} + \mathrm{CP_g} + \Delta\mathrm{LC} + \mathrm{EC} \quad (7.14)$$

where:

$$\mathrm{CRF} = \frac{i(1+i)^n}{(1+i)^n - 1} \quad (7.15)$$

and n is the number of years in the lifetime in the system of which the interest rate i is considered

$$\mathrm{COP} = \mathrm{ACS}/\sum \mathrm{TQW}(t) + \mathrm{Salt_p} \tag{7.16}$$

$$\mathrm{LCOE} = \frac{\mathrm{ACS} + P_{\mathrm{WD}}(t) + P_B(t)}{\sum P_{\mathrm{PV}}(t) + W_p(t) + \mathrm{GP}_i(t)} \tag{7.17}$$

$$\forall\, t = 1, 2, 3, \ldots 8760.$$

The optimisation problem is formulated as a multi-objective linear programming problem as expressed in (7.18), with the objective of minimising the ACS and EC while maximising the quantity of freshwater production subject to constraints expressed by (7.19)–(7.24). Weighting factors (W_1, W_2, W_3, and W_4) were assigned based on the preference of significant concern, with the global warming impacts (GWI) of emission ranking highest. The ACS and the total quantity of water produced (TQW(t)) rank lowest since meeting water demand is more critical in water production

Objective Function

$$\mathrm{Min}\left[w_1 \times \mathrm{ACS} + w_2 \times \sum \mathrm{GWI} + w_3 \times \sum \mathrm{QB}(t) - w_4 \times \sum \mathrm{TQW}(t)\right] \tag{7.18}$$

S.t.

$$P_{\mathrm{PV}}(t) + W_p(t) + \mathrm{GP}_i(t) = P_{\mathrm{WD}}(t) + P_B(t) + \Delta L(t) + \mathrm{GP}_e(t) \tag{7.19}$$

$$\mathrm{TQW}(t) \geqslant \mathrm{WD}(t) \tag{7.20}$$

$$\mathrm{QW}^{\min} \leqslant \mathrm{QW_{RO}}(t) \leqslant \mathrm{QW}^{\max} \tag{7.21}$$

$$P_{\mathrm{WD}}{}^{\min} \leqslant P_{\mathrm{WD}}(t) \leqslant P_{\mathrm{WD}}{}^{\max} \tag{7.22}$$

$$\forall\, t = 1, 2, 3, \ldots 8760$$

$$\mathrm{AWT} \geqslant 0 \tag{7.23}$$

$$\mathrm{APV} \geqslant 0. \tag{7.24}$$

Constraint (7.19) expresses the power balance that ensures that power supply from PVS, a wind generator, and the grid at any time t equals the total demand by the RO desalination and brine treatment unit and the power exported to the grid at the same time t. Constraint (7.20) is a water balance that ensures the total amount of water produced at any hour t is equal to or in excess of the water demand at that hour t, while constraint (7.21) ensures that the water produced per hour remains within the required limits. Constraint (7.22) is the limit of power required by RO to produce water at any time t, and constraint (7.23) expresses the limit of the AWT while constraint (7.24) limits the area of PVs.

Using the same metrological data from Stellenbosch University in the Western Cape province of South Africa in the previous chapter, the hourly wind speed, as represented in figure 7.2 is considered for these models, whereas figure 7.3 represents the assumed water demand of the residence for every hour of the day.

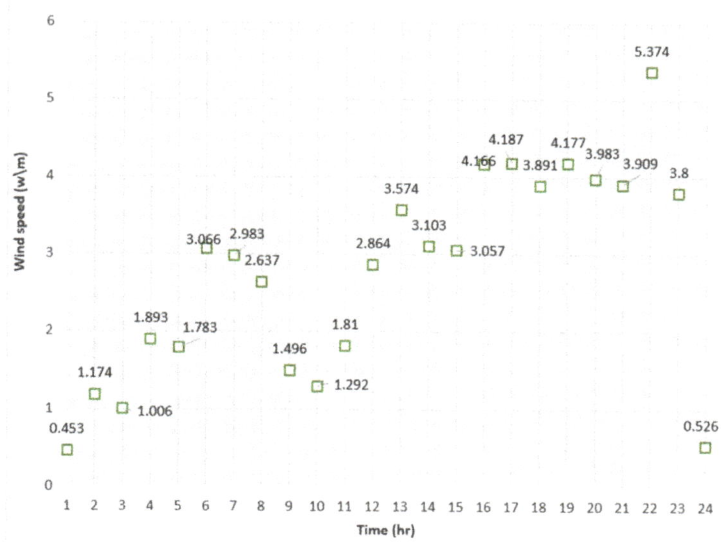

Figure 7.2. Hourly wind speed of a coastal area of Cape Town, South Africa. Reproduced from [27] CCBY 4.0.

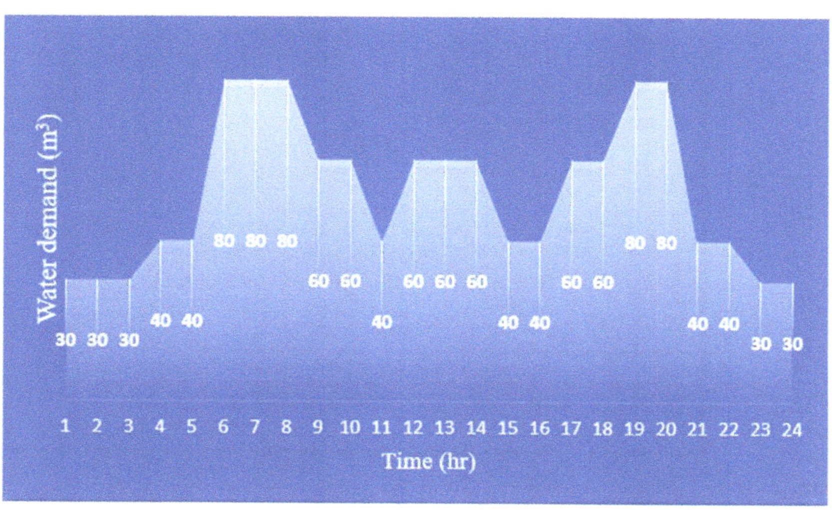

Figure 7.3. Hourly water demand by the consumers. Reproduced from [27] CCBY 4.0.

7.2.5 Results and discussion

The hourly power supply and the electricity sold back to the grid are shown in figure 7.4, together with the power generated by the wind generator and the power output generated by the PV generator. In order to keep things as simple as possible, a simulation of a typical day is used. The amount of freshwater created each hour by the three different units—RO, ED, and CRY—is seen in figure 7.5. Figure 7.6 illustrates

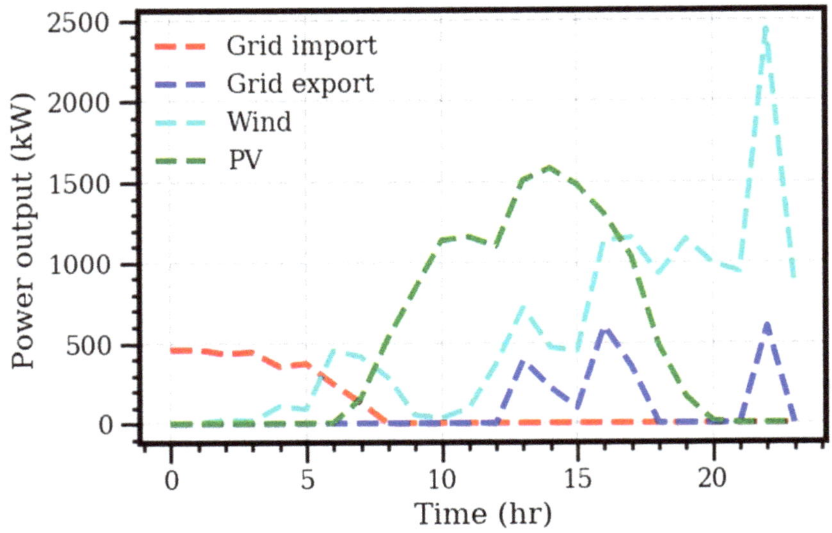

Figure 7.4. Hourly energy output from the three energy sources and the excess sold back to the grid. Reproduced with permission from [18].

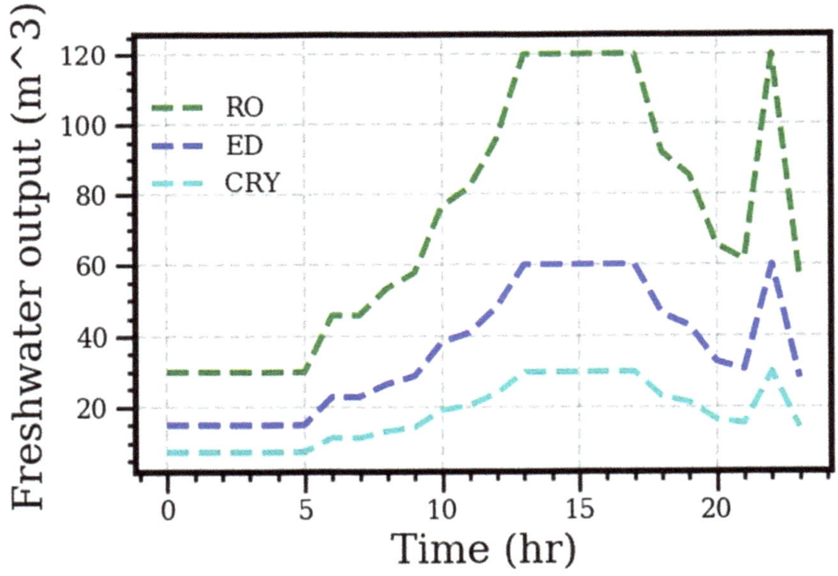

Figure 7.5. Hourly freshwater dispensed by the three segments of hybrid desalination system. Reproduced with permission from [18].

the hourly amount of brine, freshwater, and salt that is generated, while table 7.1 provides a summary of the results of the optimisation of the many other parameters.

Figure 7.4 shows the outcome of this investigation, which demonstrates the influence of ToU DR on moving demand away from peak hours to off-peak and

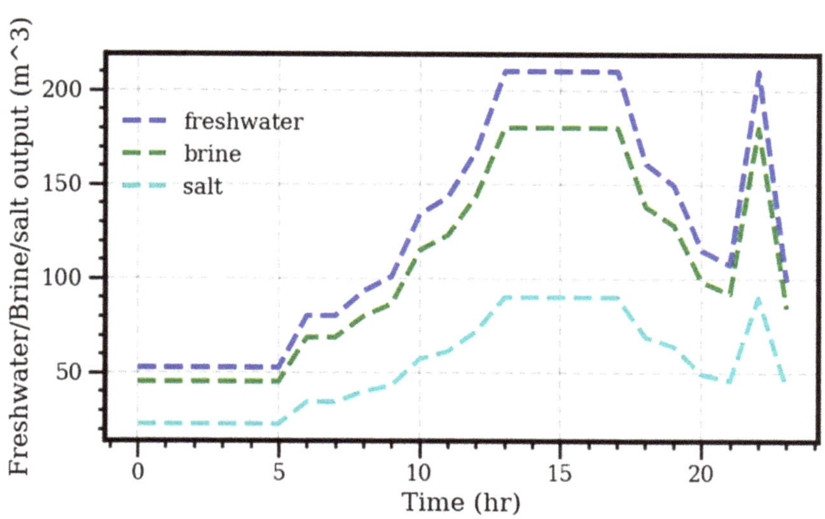

Figure 7.6. Hourly freshwater, brine, and salt produced from the integrated desalination and brine treatment unit. Reproduced with permission from [18].

Table 7.1. Summary results of optimised parameters. Reproduced with permission from [18].

ACS ($)	Emission ($kgCO_{2-e}$)	Emission cost ($)	Daily water produced (m^3)	Daily salt produced (m^3)	LCOE ($ kW^{-1})	Unit COP ($ m^{-3})	Area of PV (APV) (m^2)	AWT (m^2)
1 401 367	637 708	261 460	3005	1288	0.89	1.056	9431	50 891

normal hours for more cost-effective energy management. The selection of energy sources and the subsequent power is dependent on the cost and the EC of the energy source; this is the reason for the low power output of the grid, which is mostly dependent on generators that use fossil fuels. Because the carbon that is released by the grid generator has a cost associated with it (261 460 $ $year^{-1}$), the cost of grid power delivery has increased. Because of this, renewable energy sources such as wind and solar PV generators have an advantage, which explains the high power production from these two sources.

According to the hourly volume of freshwater generated by the three units (RO, ED, and CRY) shown in figure 7.5, the RO unit, which is the primary desalination unit, generates the maximum amount of freshwater, followed by the ED unit. The CRY unit produces the least amount of freshwater. Since the higher amount of brine delivered to the CRY unit already contains a high concentration of salt, the CRY unit only generates a tiny volume of freshwater. The CRY process is the primary unit for brine treatment because it produces soluble salts and a tiny amount of freshwater while simultaneously leaving no brine outflow.

Figure 7.6 illustrates the total amount of drinkable water generated by RO-ED-CRY at each hour of the day, together with the hourly brine output of the RO unit and the salt created from the brine itself. On the basis of these data, it seems that a significant portion of the feed water is transformed into freshwater, while the remaining portion is crystallised into salts. Additionally, the volume of freshwater produced by this combined model of brine treatment and desalination (3005 m^3/day), as shown in table 7.1, is higher than the volume produced by a standalone RO unit previously presented by reference [28], with similar input parameters. This is also higher than the expected desalination capacity to meet the baseline daily water demand (1250 m^3/day), shown in table 7.1. Furthermore, this research provides the unit COP as a single unit cost of production. This is because it is impossible to separate the COP because the system is constructed as a single unit with the same component prices, energy sources, and their respective costs. The unit COP ($0.89) is within the range reported in the literature for the COW for standalone desalination machines, which ranges from $0.5–$2.39 [28–30], respectively. Furthermore, many standalone RO desalination systems do not consider the cost of EC and brine treatment. As a result, they use less energy and treatment chemicals and require less frequent membrane replacement than a combined model that produces freshwater and treats brine while discharging nothing. Because of this, the per-unit cost of producing this combination model is quite cheap, taking into account the issues related to the economy and the environment. On the other hand, the LCOE of this model is 1.06 $ kWh^{-1}, which falls within the average value of standalone RO desalination systems, which typically vary between 0.5 and 1.2 $ kWh^{-1}. This indicates that the energy cost is comparable for standalone desalination units and combined desalination-brine treatment units. This is because the component cost of comparable energy sources utilised for a solo unit is the same for a combined unit and mostly depends on the size of the system and the amount of output that is generated.

A sensitivity analysis was performed to analyse the consequences of a percentage increase in water demand on three different cost matrices, illustrated in figures 7.7–7.9, respectively. These matrices are ACS, LCOE, and COP.

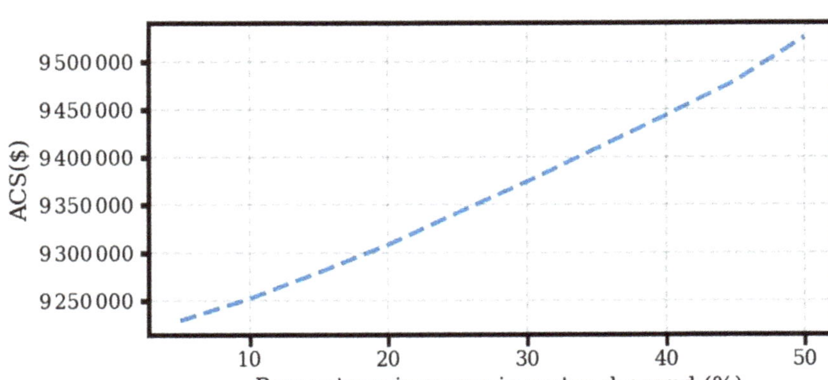

Figure 7.7. Impacts of percentage increased on the annual cost of the system. Reproduced with permission from [18].

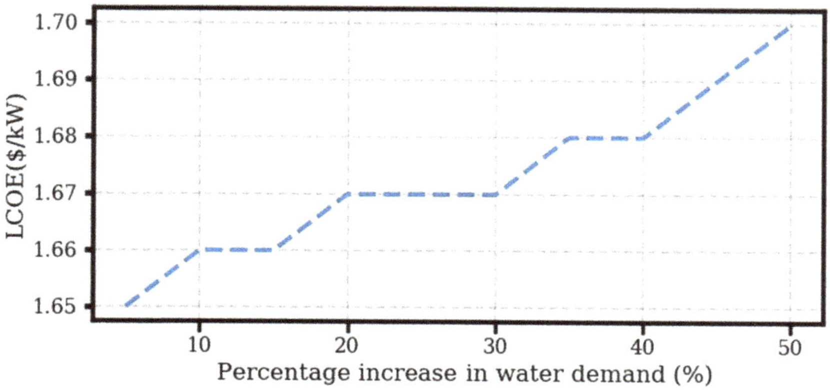

Figure 7.8. Impacts of increased percentage on LCOE. Reproduced with permission from [18].

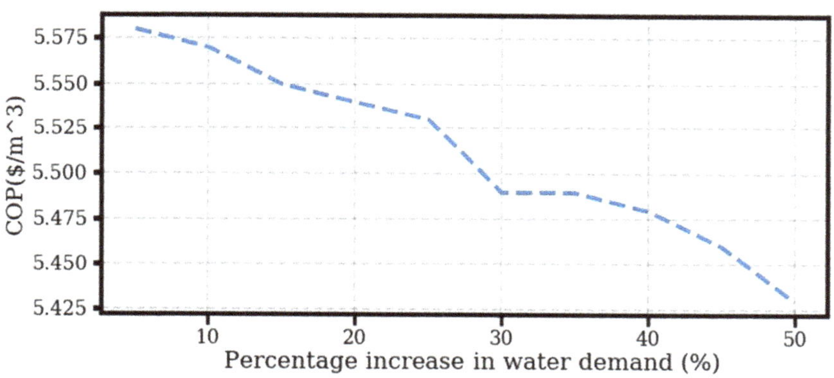

Figure 7.9. Impacts of percentage increase in the COW and salt. Reproduced with permission from [18].

The findings of the sensitivity analysis indicate that the ACS will grow proportionally, but the LCOE will increase somewhat and the cost for products will fall. These findings are in response to the percentage increase in water demand. This suggests that the growth in water demand causes an increase in both ACS and LCOE but that the unit COP falls as a consequence of increasing freshwater production to satisfy the increase in water demand.

7.3 Summary

Case study 1 in this chapter offers an effective way to manage both energy and brine throughout the process of producing freshwater through the combination of RO, ED, and CRY techniques. This model aims to reduce both the LCOE and the amount of brine produced while simultaneously increasing the amount of freshwater and salt extracted at the lowest possible price. The suggested design is such that the input water, which is salty water, is desalinated by the RO unit; the brine that is created by the RO unit is further desalinated by the ED technique, leaving behind a very high concentration that may be crystallised into soluble salts, therefore meeting the goal of producing no brine at all. Furthermore, to efficiently manage energy, an optimum sizing of energy sources was performed, such as grid power, wind power, and solar power. This was done with the mitigation of EC and its cost taken into account, as well as the intermittent restriction of renewable energy sources (RES). Due to the integrated design, both the internal and external expenses of the desalination process were analysed and reduced as much as possible. The findings demonstrate that there is a correlation between emissions and the price of energy, which drives up the price of electricity supplied by the grid and makes RES not only more environmentally benign but also more cost-effective. The LCOE is within the average value of those of standalone desalination units, suggesting that the cost of energy is comparable to a standalone desalination unit and a combined desalination-brine treatment unit. This is because the component prices of comparable energy sources utilised for a standalone unit are the same for a combined unit and mostly rely on the size of the system and the amount of output it generates. It is possible for additional research to explore the cost of salt based on location, as it is anticipated that the salt generated, if further treated, may be an additional benefit of the combined model of desalination and brine treatment, as it can be useful. Additionally, annual ECs (637 708 $kgCO_{2-e}$) are still relatively high, and a further decrease can be obtained by integrating a storage system to compensate the inconsistency of the RES. This would allow for a further reduction in ECs. This will reduce reliance on the electricity provided by the grid but may increase production costs.

7.4 Case study 2: hybrid PRO-RO desalination system using brine for salinity gradient energy production

Most current attempts to achieve sustainable desalination are focused on increasing energy efficiency and addressing environmental issues via using clean RES and the effective management of concentrated brine. Thus far, solar and wind power are mostly used for desalination and integrated with thermal or fossil fuel generators to compensate for the intermittent nature of the two. Solar and wind energy and energy storage devices have been investigated as potential complements to thermal or fossil fuel energy for desalination. This is done to lower the amount of thermal or fossil fuel energy required for desalination. To compensate for the unpredictability of solar and wind energy, research has been conducted on various energy storage devices. In areas where it is possible to do so, geothermal energy has also been studied for use in the desalination process. A substantial amount of research has not been done to investigate the entire scope and potential of exploiting ocean energy for desalination applications [31]. Research on the utilisation of wave and tidal energy has been conducted, although in small amounts, with the majority of this research still in the experimental phases. Because it can be predicted and has the maximum energy density, ocean energy is more dependable than other kinds of renewable energy [31]. Other benefits of using ocean energy for desalination include natural collocation of production and usage, which eliminates the need for and associated costs of energy transportation, as well as near proximity to human settlements since many human populations are placed around coastlines [31].

Additionally, ocean energy may be broken down into other types, including thermal, mechanical, and chemical (salinity gradient) energy. These types allow for dynamic applications because of their flexibility. Therefore, the purpose of this paper is to offer a deeper investigation into the use of one kind of ocean energy (salinity gradient) with its one-of-a-kind feature alongside wind energy with the hybrid of PRO and RO.

The desalination procedures may be classified as thermal or membrane-based, as was covered previously in chapter 2. In most cases, the membrane processes, such as PRO and RO, needed just electrical energy or pressure differences, and their SEC values ranged from 2 to 6 kWh m^{-3}. Although the membrane process requires less energy, every method of desalination has some drawbacks that are unique to that method and faces the same obstacle of brine management. For instance, RO is the most dependable membrane technology for desalinating saltwater. However, it still has issues with membrane fouling and rejecting certain soluble contaminants. An extensive study on the production of membranes via the use of nanomaterials has been prompted by the improvement of RO systems [32–34]. In addition, several studies have proposed combining RO and other desalination techniques to boost performance. The authors of [9, 18, 22] devised a model that incorporated RO, ED, and CRY processes to accomplish salt production from saltwater desalination while minimising the amount of brine that was produced. The economic study did not consider the capital and operating costs of ED and CRY, even though it analysed the costs of the systems as well as their energy needs and found them to be suitable for brine management and economically feasible for that purpose. The hybridisation of RO with either FO, reverse electro-dialysis (RED), or PRO is seen as having a great deal of potential, given that all three of these processes are significant technologies that make use of the energy provided by ocean salinity gradients. The authors of [10–12] discuss the hybrid FO-RO technologies and analyse the system's performance in addition to determining whether or not it is economically viable. It was hypothesised that the FO-RO hybrid might reach an SEC threshold of 1.3–1.5 kWh m^{-3} for the desalination of saltwater by using an FO membrane with a capacity of about 10 L m^{-3} [10]. This hybrid approach decreases the amount of energy that is required, the cost of freshwater, and the environmental effect of brine.

7.4.1 System architecture

The suggested structure of the PRO system is such that the process would come first, followed by the process of RO, which would then desalinate the seawater. At the PRO stage, ocean water is utilised as a low salinity solution to produce a salinity gradient with the concentrated draw solution (brine), which is the by-product of the desalination process. The salinity gradient is the pressure differential between the low concentrated feed water and the high concentrated brine solution that enables feed water passage through the PRO membrane. This pressure difference is caused

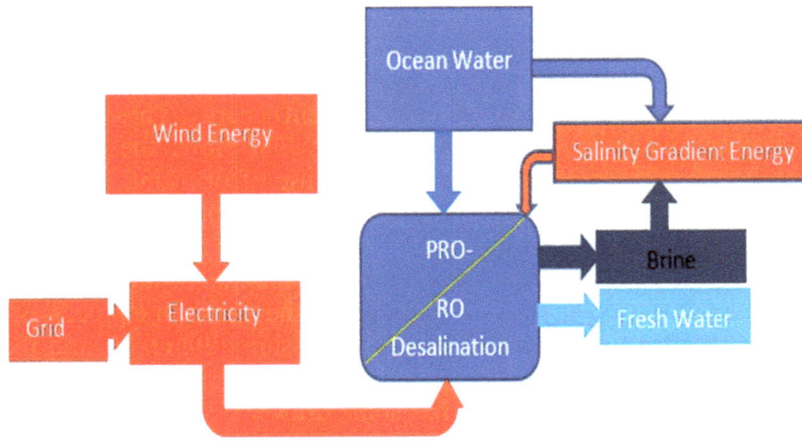

Figure 7.10. Integrated PRO-RO desalination process powered by grid-tied offshore wind and ocean salinity gradient. Reproduced from [27] CC BY 4.0.

by the difference in the concentration of the two solutions. After passing through the PRO, the input water with low salinity is sent into the RO section, which requires a low pressure to desalinate the water. Wind energy, with grid power serving as a backup, is used to provide the low pressure required for the RO unit. The integrated PRO-RO desalination process is shown in figure 7.10 as a schematic figure. This process is driven by the salinity gradient and grid-tie offshore wind.

7.4.2 System cost model

As a result of the disparity in osmotic pressure between the two solutions, the PRO process allows the passage of water over a membrane from a solution with a lower concentration to a solution with a higher concentration. Because it requires a back pressure on the high concentration side, which retards permeate flow, and it can create power when the solution is depressurised through a turbine, it is a highly promising technology for the production of energy from the salinity gradient [35, 36]. The power required by PRO is almost entirely for the pressure pump, and this is something that a salinity gradient can easily provide at no additional expense.

Both the osmotic pressure difference ($\Delta\pi$) and the hydrostatic pressure difference (ΔP) may be used to describe the propelling force that takes place across the membranes in PRO. Calculating the pressure-retarded osmotic water flux J_w that results from this can be done from the membrane water permeability A and the propelling force of movement [27, 37]:

$$J_w = A(\Delta\pi - \Delta P). \tag{7.25}$$

In the same form as hydroelectric power, the power density P_d produced in a PRO process is the product of the augmented flow rate and pressure drop through a hydro-turbine:

$$P_d = J_w \Delta P = A(\Delta \pi - \Delta P)\Delta P. \tag{7.26}$$

The differentiation of (7.26) will give the maximum power as (7.27)

$$P_{d,\,\max} = \frac{A}{f}\Delta \pi^2. \tag{7.27}$$

This is achieved when the applied force is half the osmotic pressure ($\Delta P = \Delta\pi/2$). The net salinity gradient power output (SG_P) from the PRO plant would be 'generated power' minus 'consumed power,' expressed in (7.28) [27, 38, 39]

$$SG_p = SG_{pg} - SG_{pc}. \tag{7.28}$$

It is unlikely to provide an accurate estimate for the capital cost of a PRO system given that there is currently a commercial PRO plant in operation. With the exception of experimental situations and test systems, the literature does not include any specific data on the capital cost of PRO systems; therefore, this means that such information is limited [40, 41]. As a result, we use a technique very similar to that described in [39–41], which benchmarks the RO system's capital expenditure for PRO.

The energy sources used to power this integrated desalination and brine treatment system are electricity from the traditional power grid and RES from the wind and salinity gradient. The same grid power output and cost model from chapter 6 (equation (6.1)–(6.6)) is utilised while the wind power output and its cost model discussed in case study 1 of this chapter is adopted. We shall only detail the salinity gradient energy output in this case to complete the energy source model to be sized and evaluated. Wind power is the primary source of propulsion for the RO component of the desalination plant. This designed system suggests that the wind farm is in close proximity to both the sea and the RO unit; as a result, the costs associated with power transmission and energy losses are estimated to be low.

7.4.3 Salinity gradient energy

The salinity gradient energy, like other forms of ocean energy, is very predictable and has a high energy density; therefore, it is more dependable than other forms of renewable energy [42]. If given the appropriate amount of attention, it has the potential to increase energy sustainability. Due to technological limitations, this kind of energy has not yet been converted into a form that can be used on a large basis. Very little effort has been made to examine the use of ocean energy, with most of these projects still in the experimental stage [42]. The difference in the salt content of the two different solutions is the source of this energy, which is called the blue energy. It can be captivated and transformed into electrical energy or it can be used directly for desalination using technologies such as FO, RED, and PRO. It is possible to make use of the change in entropy that occurs as a consequence of the irreversible mixing of two solutions with different concentrations of salt in order to achieve the conversion of some of the thermal energy of the fluids into electrical energy [42]. While Jia *et al* [42] discussed the many methods that can be utilised to collect the energy from salinity gradients, Tufa *et al* [37] went into more depth on the thermodynamic features of the process. Indicating that if the chemical potential μ_i of a solution is known, it can be used to calculate the amount of free energy that is released during the mixing process.

$$G = \sum \mu_i n_i \tag{7.29}$$

where n is the number of moles of each component which can be expressed in terms of molar concentration c and total volume V as:

$$n_i = c_i V. \tag{7.30}$$

The Gibbs free energy of mixing (ΔG_{mix}) of two salt solutions at different concentrations is given by the difference between the Gibbs free energy of the mixed solution (G_t) and the Gibbs free energy of initial solutions:

$$\Delta G_{\text{mix}} = G_t - (G_c + G_d) \tag{7.31}$$

where subscripts c and d refer to concentrate and dilute solutions, respectively. Combining (7.29)–(7.31), ΔG_{mix} can be calculated as:

$$\Delta G_{\text{mix}} = \sum_i (c_{i,t} V_t RT \text{In} \gamma_{i,t} x_{i,t}) - \left(c_{i,c} RT \text{In} \gamma_{i,c} x_{i,c}\right) + \left(c_{i,d} RT \text{In} \gamma_{i,d} x_{i,d}\right) \tag{7.32}$$

where R is universal gas constant, T is absolute temperature of the solution, ln is natural logarithm, γ is activity coefficient.

Theoretically, the greatest amount of energy that can be extracted during irreversible mixing of a low concentrate water with high concentrate draw is enormous, ranging from 0.75 to 14.1 kWh m^{-3} depending on the low-concentration stream [37, 42]. In practise, however, the maximum amount of energy that can be extracted is much lower. Because there is currently no widely accessible commercial technology that can capture the energy of the salinity gradient for use in large-scale desalination systems, the energy of the salinity gradient will only be used in this research to power the PRO portion.

7.4.4 Optimisation problem formulation

Like in the previous optimisation model discussed earlier, this case study also uses the ACS, LCOE, and COW cost matrix for the economic assessment of the PRO-RO desalination. ACS is the sum of the total system component cost with the CRF (7.15) a function of total initial cost of the system cost (TISC) [24–26]. COW and LCOE are the other cost matrix for evaluating the cost of energy expended and the cost of producing 1 cm^3 of water, respectively [27]

$$\text{TISC} = \text{IC}_{\text{WT}} + \text{IC}_{\text{PRO}} + \text{IC}_{\text{RO}} + \text{ICW}_{\text{TK}} \tag{7.33}$$

$$\text{TMC} = \text{AMC}_{\text{WT}} + \text{AMC}_{\text{PRO}} + \text{AMC}_{\text{RO}} \tag{7.34}$$

$$\text{ACS} = \text{TISC} \times \text{CRF} + \text{TMC} + \text{AC}_{\text{MR}} \times \text{CRF} + \text{AC}_{\text{CH}}. \tag{7.35}$$

TMC is the total maintenance and operation cost, AC_{MR} and AC_{CH} are the annual cost of the membrane and treatment chemicals, respectively [27]

$$\mathrm{COW} = \mathrm{ACS}/\sum \mathrm{DQW}(t) \tag{7.36}$$

$$\mathrm{LCOE} = \frac{\mathrm{ACS}}{\sum W_p(t) + \mathrm{SG_p}(t)} \tag{7.37}$$

$$\forall\, t = 1, 2, 3, \ldots 8760.$$

The multi-objective optimisation problem is to minimise ACS while the quantity of freshwater produce is maximised as expressed by (7.38).

$$\mathrm{Min}\left[w_1 \times \mathrm{ACS} + w_2 \times \sum \mathrm{DQW}(t)\right] \tag{7.38}$$

S.t.

$$W_p(t) + G_{\mathrm{pi}}(t) = P_{\mathrm{WD}}(t) + G_{\mathrm{pe}}(t) \tag{7.39}$$

with weighting factors (W_1 and W_2), which were allocated to rank the objective components on a preferential order of significance. All other constraints are equivalent to those of case 1 ((7.20)–(7.21)).

The energy security will be evaluated as part of the reliability assessment in order to determine the overall level of robustness. In light of the unpredictable nature of wind energy, this is done to ensure that energy efficiency is maintained. This reliability evaluation centres its attention on the interruptions that are brought on by the intermittent nature of wind energy for the purpose of this research. The energy provided by the salinity gradient helps compensate for the wind's declining demand for available energy. As a result, the LOLP index is used to conduct this evaluation. LOLP is often used by utilities to determine the total number of days in a given year when the capacity of the generators could not satisfy the daily peak demand. It may also be stated as (7.40) [43]

$$\mathrm{LOLP} = \sum_{j=1}^{j} p_j \ell_j \tag{7.40}$$

where j is the capacity outage state, p_j is the state probability of the capacity outage state j, and ℓ_j is the outage duration of the capacity outage state j.

Figure 7.11 represents the flow chart of the optimisation model.

7.4.5 Results and discussion

EC cost, emission factor using the same metrological data of energy price as those in the previous case study 1, and other input parameters for optimal cost analysis are contained in table 7.2.

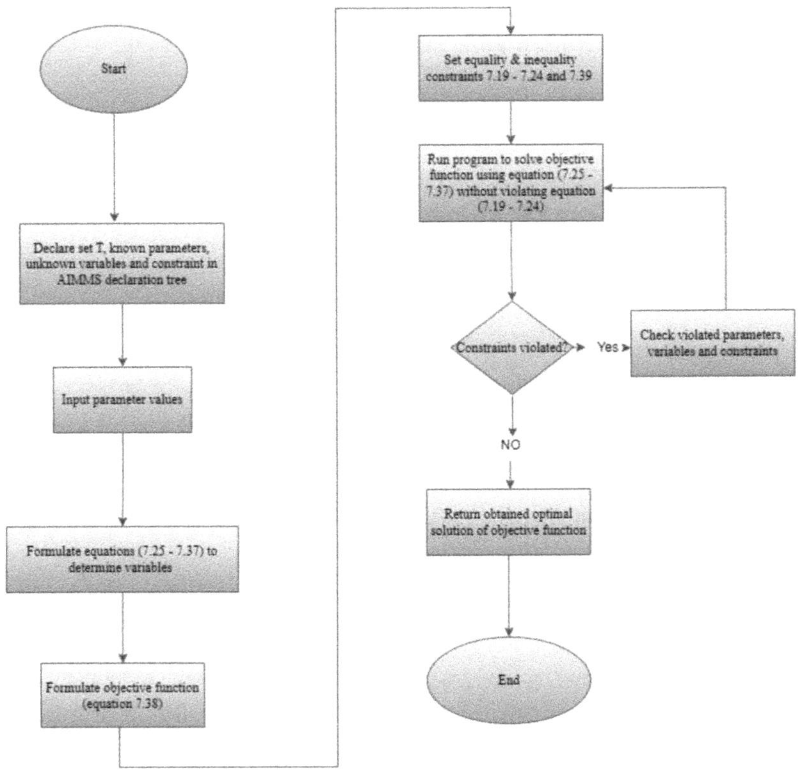

Figure 7.11. Flow chart of the optimisation model.

Table 7.2. Input parameters [18, 27, 29, 30]. Reproduced from [27] CC BY 4.0.

Parameters	Value	Parameters	Value	Parameters	Value
Project time	20 years	C_p	0.59	B	0.85 L $m^{-2}h^{-1}$
Interest rate	5%	C_{WT}	\$1804	$\Delta\pi$	25.35 bar
Discount rate	3%	C_{RO}	532\$ m^{-3} day^{-1}	A	4.0 L $m^{-2}h^{-1}$
Emission factor	1.07	Plant maintenance cost	0.2 \$ m^{-3}	WT rated power	1 kW
Emission tax	\$0.41	Chemical cost	0.6 \$ m^{-3}	C_{MR}	0.6 \$ m^{-3}
η_w	85%	CW_{TK}	255 \$ m^{-3}	ρ_{air}	1.23 kgm^3

This proposed model shows the hourly power output from the wind and grid sources as depicted in figure 7.12 and the optimal hourly quantity of water produced represented in figure 7.13. Furthermore, the summary of optimised parameters is contained in table 7.3.

Wind energy is obviously favoured over grid electricity in figure 7.12. When wind power is available, the wind generator powers the RO desalination system. When

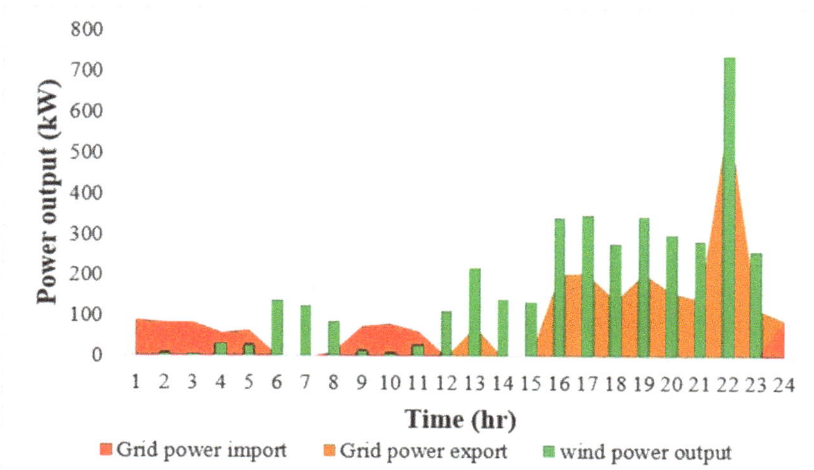

Figure 7.12. Hourly energy dispensed from the grid and wind power generators. Reproduced from [27] CC BY 4.0.

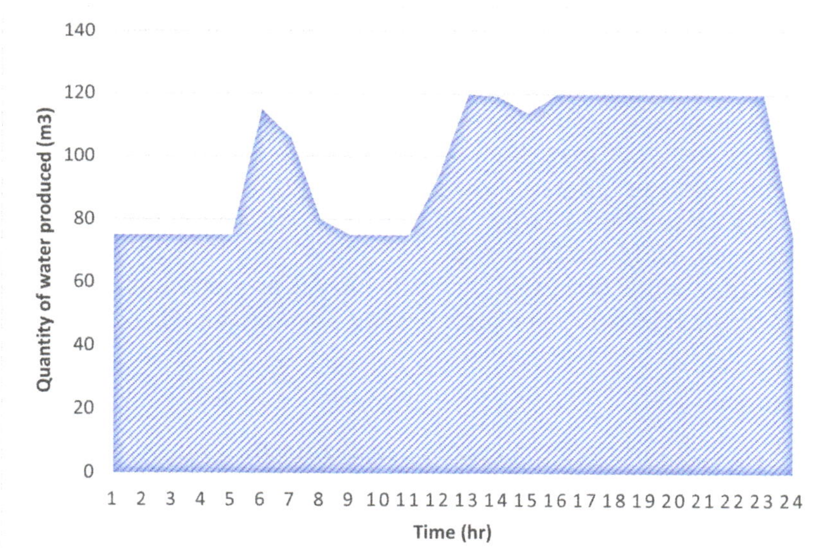

Figure 7.13. Hourly freshwater produced from the integrated PRO-RO system. Reproduced from [27] CC BY 4.0.

Table 7.3. Summary of optimised parameters. Reproduced from [27] CC BY 4.0.

ACS ($)	LCOE ($ kW^{-1})	COW ($ m^{-3})	Emission ($kgCO_{2-e}$)	Emission cost ($)	Daily water produced (m^3)	AWT (m^2)	J_W ($m^3m^{-2}h^{-1}$)
110 456	1.11	0.13	269 405	110 456	2384	15 543	17.031

there is a lack of wind energy due to wind speed fluctuation, grid electricity is a backup power source for the RO desalination machine. The surplus electricity the wind turbine generates is exported back to the grid at the same cost as buying from the grid. As mentioned in section 7.4.3, the salinity gradient energy is mostly utilised to power the PRO unit's pressure pump, allowing water to pass over the membrane and into the RO unit. This initial step might be considered a pretreatment procedure since it helps minimise the energy demand of the RO unit and lowers fouling effects on the RO membrane, thereby boosting permissibility. The result is a huge amount of water generated at every hour of the day to satisfy demand and have extra, particularly during strong wind energy production hours, as illustrated in figure 7.13. The ACS of the integrated system is \$110 456, the LCOE is 1.11 \$ kW^{-1}, and the COW is 0.13 \$ m^{-3}, according to the summary findings of the optimised parameters provided in table 7.3. These costs are lower than those in the literature for standalone RO units and comparable hybrid scenarios that did not account for EC costs and expected the component cost of PRO to be the same as that of RO in this research. For instance, [5, 40] reported 1.33 and 1.77 \$ m^{-3} for COW for a standalone RO unit, respectively. It should be emphasised that these economic figures are liable to alter depending on the real cost of the PRO component if it is known.

Furthermore, when the cost of brine disposal and management is included, the suggested approach is significantly more cost-effective since there is no need for the additional external expense of brine disposal or management. The carbon output of 269 405 $kgCO_{2-e}$ is lower than that of a comparable system using diesel generators or desalination systems primarily fuelled by the grid's traditional energy source because the greater wind energy production is sufficient to operate the RO desalination plant, and the LOLP is as low as 0.341. This is particularly true since the unit's energy need is decreased due to the lower pressure required to send feed water through the RO membrane after passing through the PRO unit.

7.5 Summary

The case study 2 of this chapter presents a model for the optimal techno-economic and reliability evaluation of a PRO-RO hybrid desalination system powered by a grid-connected wind energy source and salinity gradient energy. The design comprises two phases. The first stage is the PRO unit, powered by the salinity gradient energy generated by the combination of ocean feed water (low concentration) and RO brine solution (high concentration). In the second stage, a RO unit powered by grid-connected wind energy desalinates secondary ocean input water. The brine produced by the second stage is redirected to the first stage to generate a salinity gradient for water transmission through the PRO membrane and energy generation. The proposed model sized the grid-connected wind energy sources optimally to minimise ECs, its cost, and the entire system's ACS, LCOE, and COW. The proposed design model's tech-economic evaluation yielded an ACS of \$110 456, an LCOE of 1.11 kW m^{-3}, and a COW of 0.13 \$ m^{-3}. These costs are significantly less than the literature-available, standalone RO desalination system. For instance,

[5, 40] reported COW costs of 1.33 and 1.77 \$ m^{-3} for a solitary RO unit, respectively. The hybrid PRO-RO model is more environmentally favourable due to its extended benefits of zero saline disposal and management costs, minimal ECs, and emission costs. The PRO-RO desalination system can also be considered cost-effective, as saline disposal and treatment costs will no longer be a factor. The low LOLP reliability index of 0.341 suggests the proposed system is reliable due to the high power output from the wind energy source and the low energy required by the RO unit after the PRO unit partially desalinates the feed water.

In conclusion, the hybrid PRO-RO desalination system propelled by wind energy is a promising method for resolving freshwater deficiency and the associated environmental issues of saline disposal. In addition, it addresses the problem of high ECs caused by using a conventional fossil fuel generator for RO desalination. Future advancements in PRO technology will allow salinity gradient energy to be utilised to satisfy the energy requirements of RO desalination. This will significantly reduce the cost of energy for desalination systems. Consequently, it will minimise the cost of producing freshwater.

References

[1] Molinos-Senante M and González D 2019 Evaluation of the economics of desalination by integrating greenhouse gas emission costs: an empirical application for Chile *Renew. Energy* **133** 1327–37

[2] Baumgärtner N, Delorme R, Hennen M and Bardow A 2019 Design of low-carbon utility systems : exploiting time-dependent grid emissions for climate-friendly demand-side management *Appl. Energy* **247** 755–65

[3] Okampo E J and Nwulu N 2021 Techno-economic evaluation of reverse osmosis desalination system considering emission cost and demand response *Sustain. Energy Technol. Assess.* **46** 101252

[4] Hossam-Eldin A, El-Nashar A M and Ismaiel A 2012 Investigation into economical desalination using optimized hybrid renewable energy system *Int. J. Electr. Power Energy Syst.* **43** 1393–400

[5] Koutroulis E and Kolokotsa D 2010 Design optimization of desalination systems power-supplied by PV and W/G energy sources *Desalination* **258** 171–81

[6] Subramani A and Jacangelo J G 2014 Treatment technologies for reverse osmosis concentrate volume minimization: a review *Sep. Purif. Technol.* **122** 472–89

[7] Giwa A, Dufour V, Al Marzooqi F, Al Kaabi M and Hasan S W 2017 Brine management methods: recent innovations and current status *Desalination* **407** 1–23

[8] Panagopoulos A, Haralambous K and Loizidou M 2019 Desalination brine disposal methods and treatment technologies—a review *Sci. Total Environ.* **693** 133545

[9] Nayar K G, Fernandes J, McGovern R K, Al-Anzi B S and Lienhard J H 2019 Cost and energy needs of RO-ED-crystallizer systems for zero brine discharge seawater desalination *Desalination* **457** 115–32

[10] Yangali-Quintanilla V, Li Z, Valladares R, Li Q and Amy G 2011 Indirect desalination of Red Sea water with forward osmosis and low-pressure reverse osmosis for water reuse *Desalination* **280** 160–6

[11] Valladares Linares R, Li Z, Abu-Ghdaib M, Wei C H, Amy G and Vrouwenvelder J S 2013 Water harvesting from municipal wastewater via osmotic gradient: an evaluation of process performance *J. Memb. Sci.* **447** 50–6

[12] Valladares Linares R *et al* 2016 Life cycle cost of a hybrid forward osmosis—low-pressure reverse osmosis system for seawater desalination and wastewater recovery *Water Res.* **88** 225–34

[13] Panagopoulos A, Loizidou M and Haralambous K J 2020 Stainless steel in thermal desalination and brine treatment: current status and prospects *Met. Mater. Int.* **26** 1463–82

[14] Panagopoulos A and Haralambous K 2020 Environmental impacts of desalination and brine treatment—challenges and mitigation measures *Mar. Pollut. Bull.* **161** 111773

[15] Roberts D A, Johnston E L and Knott N A 2010 Impacts of desalination plant discharges on the marine environment: a critical review of published studies *Water Res.* **44** 5117–28

[16] Lior N 2017 Sustainability as the quantitative norm for water desalination impacts *Desalination* **401** 99–111

[17] Mannan M, Alhaj M, Mabrouk A N and Al-Ghamdi S G 2019 Examining the life-cycle environmental impacts of desalination: a case study in the State of Qatar *Desalination* **452** 238–46

[18] Okampo E J and Nwulu N 2020 Optimal design and techno-economic evaluation of renewable energy powered combined reverse osmosis desalination and brine treatment unit *Desalin. Water Treat.* **202** 27–37

[19] Jones E, Qadir M, van Vliet M T H, Smakhtin V and mu Kang S 2019 The state of desalination and brine production: a global outlook *Sci. Total Environ.* **657** 1343–56

[20] Nwulu N I and Xia X 2017 Optimal dispatch for a microgrid incorporating renewables and demand response *Renew. Energy* **101** 16–28

[21] Salameh Z M and Borowy B S 1996 Methodology for optimally sizing the combination of a battery bank and PV array in a wind/PV hybrid system *IEEE Trans. Energy Convers.* **11** 367–75

[22] Nayar K G *et al* 2019 Cost and energy requirements of hybrid RO and ED brine concentration systems for salt production *Desalination* **456** 97–120

[23] Dehnavi E and Abdi H 2016 Optimal pricing in time of use demand response by integrating with dynamic economic dispatch problem *Energy* **109** 1086–94

[24] Acuña L G, Lake M, Padilla R V, Lim Y Y, Ponzón E G and Soo Too Y C 2018 Modelling autonomous hybrid photovoltaic-wind energy systems under a new reliability approach *Energy Convers. Manag.* **172** 357–69

[25] Al-Obaidi M A, Filippini G, Manenti F and Mujtaba I M 2019 Cost evaluation and optimisation of hybrid multi-effect distillation and reverse osmosis system for seawater desalination *Desalination* **456** 136–49

[26] Koroneos C, Dompros A and Roumbas G 2007 Renewable energy driven desalination systems modelling *J. Clean. Prod.* **15** 449–64

[27] Okampo E J and Nwulu N 2022 Economic and reliability assessment of hybrid PRO-RO desalination systems using brine for salinity gradient energy production *Sustainability* **14** 3328

[28] Okampo E J and Nwulu N I 2020 Optimal energy mix for a reverse osmosis desalination unit considering demand response osmosis *J. Eng. Des. Technol.* **18** 1287–303

[29] Abdelshafy A M, Hassan H and Jurasz J 2018 Optimal design of a grid-connected desalination plant powered by renewable energy resources using a hybrid PSO–GWO approach *Energy Convers. Manag.* **173** 31–47

[30] Wu B, Maleki A, Pourfayaz F and Rosen M A 2018 Optimal design of stand-alone reverse osmosis desalination driven by a photovoltaic and diesel generator hybrid system *Sol. Energy* **163** 91–103

[31] Li Z, Siddiqi A, Anadon L D and Narayanamurti V 2018 Towards sustainability in water-energy nexus: ocean energy for seawater desalination *Renew. Sustain. Energy Rev.* **82** 3833–47

[32] Teow Y H and Mohammad A W 2019 New generation nanomaterials for water desalination: a review *Desalination* **451** 2–17

[33] Mamah S C, Goh P S, Ismail A F, Suzaimi N D, Ahmad N A and Lee W J 2021 Flux enhancement in reverse osmosis membranes induced by synergistic effect of incorporated palygorskite/chitin hybrid nanomaterial *J. Environ. Chem. Eng.* **9** 105432

[34] Ng Z C, Lau W J, Matsuura T and Ismail A F 2021 Thin film nanocomposite RO membranes: review on fabrication techniques and impacts of nanofiller characteristics on membrane properties *Chem. Eng. Res. Des.* **165** 81–105

[35] Kim Y C and Elimelech M 2013 Potential of osmotic power generation by pressure retarded osmosis using seawater as feed solution: analysis and experiments *J. Memb. Sci.* **429** 330–7

[36] Long R, Lai X, Liu Z and Liu W 2019 Pressure retarded osmosis: operating in a compromise between power density and energy efficiency *Energy* **172** 592–8

[37] Tufa R A, Curcio E, Fontananova E and Di Profio G 2017 Membrane-based processes for sustainable power generation using water: pressure-retarded osmosis (PRO), reverse electrodialysis (RED), and capacitive mixing (CAPMIX) *Comprehensive Membrane Science and Engineering* 2nd edn (New York: Elsevier Science) pp 206–48

[38] Sakai H *et al* 2016 Energy recovery by PRO in seawater desalination plant *Desalination* **389** 52–7

[39] Matsuyama K *et al* 2021 Power generation system based on pressure retarded osmosis with a commercially-available hollow fibre PRO membrane module using seawater and freshwater *Desalination* **499** 114805

[40] Makabe R, Ueyama T, Sakai H and Tanioka A 2021 Commercial pressure retarded osmosis systems for seawater desalination plants *Membranes (Basel)* **11** 69

[41] Loeb S 2002 Large-scale power production by pressure-retarded osmosis, using river water and sea water passing through spiral modules *Desalination* **143** 115–22

[42] Jia Z, Wang B, Song S and Fan Y 2014 Blue energy: current technologies for sustainable power generation from water salinity gradient *Renew. Sustain. Energy Rev.* **31** 91–100

[43] Gbadamosi S L and Nwulu N I 2020 Optimal power dispatch and reliability analysis of hybrid CHP-PV-wind systems in farming applications *Sustain* **12** 8199

Part V

Conclusion and policy framework

Chapter 8

General conclusion

This chapter draws a general conclusion of the book, highlighting prospects, challenges, and present policy frameworks for sustainable energy–water nexus. The chapter identifies technological advancement, renewable energy integration, desalination-integrated water management, decentralisation and modularisation, resource recovery and circular economy, innovation in brine management, and desalination in agriculture as prospects of energy–water integrated systems. To tackle identified challenges such as data availability, technological complexity, and the host of others faced by energy–water systems, a complimentary policy framework is formulated.

8.1 Conclusion

The book conceptualises desalination techniques and focuses particularly on the most common method—reverse osmosis desalination technique. It explores the merit of this technique and its hybridisation with other techniques to achieve optimal freshwater production at minimal cost and environmental impact. It also presents different energy mix scenarios for efficient energy management needed for water treatment. This led to the design of optimal models with renewable energy sources in the energy mix to reduce the environmental pollution of carbon emission from conventional energy sources while the intermittent limitation of some of the renewable energy sources was complemented. In addition to grid energy, two common forms of renewable energy sources (RES) (photovoltaics and wind energy) were explored to power the reverse osmosis (RO) desalination system. Time-of-use (ToU) demand response (DR) was implemented to obtain the most energy-efficient schedule.

Additionally, the integration of RO, electro-dialysis, and crystallisation desalination methods was investigated to maximise freshwater production and reduce brine output. The integrated modelled systems' techno-economic assessments were

doi:10.1088/978-0-7503-6154-5ch8

presented; the findings were contrasted, and sensitivity analysis was undertaken. According to the findings, for best performance in terms of economic cost and environmental effect, the RO desalination system should be better integrated with other techniques of desalination and powered by a combination of grid electricity and RES. The use of the ToU DR programme increased production at a lower cost.

Furthermore, an integrated pressure-retarded osmosis (PRO)-RO desalination system was modelled to co-generate freshwater and salinity gradient energy. This system utilised the brine produced from the RO unit to generate the salinity gradient energy needed by the PRO. Therefore, there is no extra cost of brine treatment, and its effect on the environment is also avoided.

Currently, techno-economic analysis of RES-RO desalination has been a major research topic; however, optimisation of thermodynamic parameters such as exergy destruction and specific energy consumption has received little attention. In the future, the techno-economic evaluation of a desalination system should detail every section of the system, from sizing to operation and thermodynamic processes. Furthermore, optimisation aims in this field have been increasingly focused on economic performance metrics. A more comprehensive study should be done in the future to integrate both an economic and a reliability index in formulating objective functions.

8.2 Future outlook

The future outlook of desalination systems holds great promise and potential as the world grapples with increasing water scarcity, growing population, and climate change. Advancements in technology, research, and innovation are driving the evolution of desalination, making it a vital component of sustainable water management strategies. The energy–water nexus is critical to achieving a sustainable society. Understanding and managing this complex relationship is crucial for sustainable resource management, environmental protection, and economic development [1]. Several key trends and developments shape the future of desalination systems.

- Technological advancements: desalination technologies advance rapidly, improving energy efficiency, reducing costs, and enhancing overall performance. Innovative membrane materials, high-pressure pumps, energy recovery devices, and intelligent control systems are being developed to optimise desalination processes [2]. Emerging technologies, such as forward osmosis and membrane distillation, also show great potential to revolutionise the desalination landscape.
- Renewable energy integration: to address environmental concerns and minimise the carbon footprint of desalination, integrating RES has gained significant traction [3, 4]. Solar, wind, tidal, and wave energy are increasingly being combined with desalination systems to power plants sustainably [5]. Hybrid renewable energy systems and energy storage solutions ensure continuous operation, making desalination more environmentally friendly and economically viable [6].
- Desalination-integrated water management: desalination is transitioning from a standalone solution to an integral part of comprehensive water management strategies. Water-stressed regions adopt integrated water management

approaches, combining desalination with water reuse, rainwater harvesting, and efficient water distribution systems. This integrated approach helps optimise water resources, reduce environmental impacts, and enhance water security [7].

- Decentralisation and modularisation: decentralised desalination plants and modular systems are gaining popularity due to their flexibility and scalability [8]. Smaller-scale desalination units can be deployed closer to the point of demand, reducing transmission losses and infrastructure costs. These decentralised systems offer opportunities for community-level water supply and emergency response during natural disasters.
- Resource recovery and circular economy: desalination and water reuse are evolving from a single-purpose process to a resource-recovery and circular economy model [9]. The focus is shifting towards maximising resource utilisation by extracting valuable components from brine and seawater, such as minerals, nutrients, metals, and energy [10, 11]. This approach aligns with the principles of sustainability and resource efficiency [12].
- Innovation in brine management: efforts to tackle the environmental impact of brine discharge are intensifying. Research is being conducted on innovative brine management techniques, including enhanced dilution, brine reinjection, and brine mineral extraction [13, 14]. By minimising the environmental impact of brine disposal, desalination systems can become more sustainable and socially acceptable [15].
- Desalination in agriculture: agriculture is a major water consumer, and the integration of desalination in agriculture is gaining momentum [16, 17]. Brackish water desalination and desalination-powered irrigation systems can enhance agricultural productivity while conserving freshwater resources for other essential needs.

As the water-energy nexus gains recognition, the integration of desalination with renewable energy, energy storage, and smart grid technologies becomes essential. This integration optimises energy consumption and enhances desalination processes' overall efficiency and sustainability [18]. The future of desalination systems is characterised by technological innovations, increased sustainability, and integration with broader water management strategies. Advancements in membrane technology, renewable energy integration, resource recovery, and brine management are transforming desalination from an energy-intensive process to a sustainable and indispensable solution for addressing global water challenges. By embracing these trends and adopting holistic approaches, desalination will be vital in ensuring water security and resilience in the face of an increasingly uncertain water future [19].

8.3 Challenges faced by energy–water system optimisation

Energy–water system optimisation faces several challenges due to energy and water systems' complex and interconnected nature. Addressing these challenges is essential to achieving sustainable resource management and enhancing the efficiency and resilience of the energy–water nexus. Some of the key challenges include:

(a) Interdependency and trade-offs: energy and water systems have interdependent relationships where changes in one system directly affect the other [20]. For example, water-intensive energy production processes impact water availability while energy-intensive water extraction and treatment affect energy consumption. Optimising one aspect may lead to unintended consequences or trade-offs in the other system, requiring a careful balancing act [21].
(b) Data availability and quality: energy–water system optimisation relies on accurate and comprehensive data on energy demand, water availability, infrastructure, and resource characteristics. However, data collection can be challenging, particularly in developing regions or decentralised systems. Inconsistent or low-quality data can limit the accuracy and reliability of optimisation models.
(c) Technological complexity: energy–water system optimisation involves complex mathematical models and simulations, considering various variables, constraints, and uncertainties [22]. Implementing these models often requires specialised expertise and computational resources, making it a challenging task for some stakeholders and regions.
(d) Infrastructure constraints: existing infrastructure, such as water supply networks, energy grids, and desalination plants, may not be designed for optimal interconnection and efficiency. Retrofitting or upgrading infrastructure to accommodate optimisation measures can be costly and time-consuming.
(e) Policy and regulatory barriers: diverse policy frameworks and regulatory barriers can hinder the integration and coordination of energy and water systems. Inconsistent policies, conflicting regulations, and fragmented governance structures may impede the adoption of optimised solutions.
(f) Financial and economic considerations: implementing energy–water system optimisation may require significant up-front investments in infrastructure, technology upgrades, and research. Optimisation measures' financial viability and cost-effectiveness need to be carefully assessed to justify long-term benefits.
(g) Climate change and uncertainty: climate change introduces additional uncertainty to energy and water systems, affecting resource availability and demand patterns. Modelling future scenarios becomes challenging, and optimisation strategies must consider the potential impacts of climate change [23].
(h) Stakeholder engagement and collaboration: successful energy–water system optimisation requires collaboration and engagement among diverse stakeholders, including government agencies, utility companies, industries, communities, and researchers. Building consensus and cooperation can be challenging due to competing interests and priorities [24].
(i) Behavioural and cultural factors: human behaviour and cultural practices significantly influence energy and water consumption patterns. Overcoming deeply ingrained behaviours and attitudes towards resource use may require targeted educational and awareness-raising campaigns.

(j) Environmental considerations: while optimising the energy–water system, it is crucial to address potential environmental impacts such as water pollution, habitat disruption, and greenhouse gas emissions. Striking a balance between resource optimisation and environmental protection is essential for sustainable outcomes [25].

Overcoming these challenges requires a multidisciplinary approach involving engineering, policy-making, economics, and social sciences. Collaboration between stakeholders, technological advancements, supportive policies, and public awareness are key to achieving effective energy–water system optimisation for a sustainable and resilient future.

8.4 Policy framework for a sustainable energy–water nexus

A comprehensive policy framework for a sustainable energy–water nexus is essential for addressing the challenges of increasing water scarcity, growing energy demand, and climate change. By adopting an integrated and collaborative approach supported by technological innovation and sound governance, societies can achieve a harmonious balance between energy and water resources, ensuring a resilient and sustainable future for future generations. The following policy roadmap is recommended for adoption by government and industry policymakers.

(a) Integrated resource management:
  1. Establish a national or regional integrated resource management committee to oversee the energy–water nexus and coordinate policy development across relevant sectors.
  2. Encourage collaboration between energy and water regulatory agencies to foster holistic decision-making and optimise resource utilisation.

(b) Data collection and sharing:
  1. Implement a standardised data collection and reporting system for energy and water resources, including consumption, production, and infrastructure data.
  2. Promote data sharing between government agencies, research institutions, and industry stakeholders to facilitate accurate modelling and optimisation efforts.

(c) Technological innovation and research:
  1. Invest in research and development to advance energy and water efficiency technologies, optimisation models, and renewable energy integration.
  2. Offer grants and incentives to support innovative projects demonstrating sustainable energy–water solutions' feasibility.

(d) Infrastructure upgrades and modernisation:
  1. Develop a long-term infrastructure improvement plan to retrofit existing energy and water infrastructure for enhanced efficiency and interconnection.

   2. Prioritise investment in smart grids, water recycling systems, and advanced desalination technologies to minimise resource waste.
- (e) Renewable energy integration:
   1. Set renewable energy integration targets for desalination and water treatment plants, incentivising solar, wind, and other sustainable energy sources.
   2. Encourage public-private partnerships to develop hybrid renewable energy systems for continuous and reliable energy supply to water facilities.
- (f) Energy–water efficiency standards:
   1. Establish energy and water efficiency standards for industrial processes, power generation, and water management systems to optimise resource consumption.
   2. Provide tax incentives or rebates for industries and utilities exceeding efficiency benchmarks.
- (g) Climate resilience and adaptation:
   1. Integrate climate change considerations into energy–water planning, accounting for changing water availability, extreme weather events, and rising sea levels.
   2. Develop adaptation strategies and infrastructure resilience measures to ensure the continuity of energy and water services under changing climatic conditions [23].
- (h) Policy coherence and alignment:
   1. Review existing energy and water policies to identify conflicts or gaps and ensure policy coherence in promoting the sustainable energy–water nexus.
   2. Align the energy–water policy framework with broader sustainability, climate, and environmental policies to create synergies in resource management.
- (i) Stakeholder engagement and public awareness [24]:
   1. Facilitate stakeholder consultations with communities, industries, NGOs, and academia to gain input on policy development and foster collaboration.
   2. Conduct public awareness campaigns to educate citizens about the importance of the energy–water nexus and encourage responsible resource use.
- (j) Monitoring, evaluation, and reporting:
   1. Implement a robust monitoring and evaluation system to track the progress and effectiveness of energy–water policies and initiatives.
   2. Publish regular reports on key performance indicators, policy outcomes, and lessons learned to promote transparency and accountability.
- (k) International cooperation and knowledge exchange:
   1. Engage in international cooperation and knowledge exchange with countries facing similar energy–water challenges to share best practices and lessons learned.

2. Participate in global initiatives to address transboundary energy–water issues and promote sustainable resource management on a global scale.

References

[1] Dai J *et al* 2018 Water-energy nexus: a review of methods and tools for macro-assessment *Appl. Energy* **210** 393–408

[2] Shokri A and Sanavi Fard M 2023 Water-energy nexus: cutting edge water desalination technologies and hybridized renewable-assisted systems; challenges and future roadmaps *Sustain. Energy Technol. Assess.* **57** 103173

[3] Okampo E J and Nwulu N 2020 Optimal energy mix for a reverse osmosis desalination unit considering demand response *J. Eng. Des. Technol.* **18** 1287–303

[4] Okampo E J and Nwulu N 2021 Techno-economic evaluation of reverse osmosis desalination system considering emission cost and demand response *Sustain. Energy Technol. Assess.* **46** 101252

[5] Ashoor B B, Mansour S, Giwa A, Dufour V and Hasan S W 2016 Principles and applications of direct contact membrane distillation (DCMD): a comprehensive review *Desalination* **398** 222–46

[6] Suwaileh W, Zargar M, Abdala A, Siddiqui F A, Khiadani M and Abdel-Wahab A 2022 Concentration polarization controlin stand-alone and hybrid forward osmosis systems: recent technological advancements and future directions *Chem. Eng. Res. Des.* **178** 199–223

[7] Paul N and Elango L 2018 Predicting future water supply-demand gap with a new reservoir, desalination plant and waste water reuse by water evaluation and planning model for Chennai megacity, India *Groundw. Sustain. Dev.* **7** 8–19

[8] Amaral K D S and Navoni J A 2023 Desalination in rural communities of the Brazilian semi-arid region: potential use of brackish concentrate in local productive activities *Process Saf. Environ. Prot.* **169** 61–70

[9] Mannina G, Gulhan H and Ni B J 2022 Water reuse from wastewater treatment: the transition towards circular economy in the water sector *Bioresour. Technol.* **363** 127951

[10] Cosenza A, Gulhan H, Maida C M and Mannina G 2022 Nutrient recovery from wastewater treatment by ultrafiltration membrane for water reuse in view of a circular economy perspective *Bioresour. Technol.* **363** 127929

[11] Pérez G, Gómez P, Ortiz I and Urtiaga A 2022 Techno-economic assessment of a membrane-based wastewater reclamation process *Desalination* **522** 115409

[12] Samberger C 2022 The role of water circularity in the food-water-energy nexus and climate change mitigation *Energy Nexus* **6** 100061

[13] Okampo E J and Nwulu N 2022 Economic and reliability assessment of hybrid PRO-RO desalination systems using brine for salinity gradient energy production *Sustainability* **14** 3328

[14] Panagopoulos A and Giannika V 2022 Comparative techno-economic and environmental analysis of minimal liquid discharge (MLD) and zero liquid discharge (ZLD) desalination systems for seawater brine treatment and valorization *Sustain. Energy Technol. Assess.* **53** 102477

[15] Okampo E J and Nwulu N 2020 Optimal design and techno-economic evaluation of renewable energy powered combined reverse osmosis desalination and brine treatment unit *Desalin. Water Treat.* **202** 27–37

[16] Zolghadr-Asli B, McIntyre N, Djordjevic S, Farmani R and Pagliero L 2023 The sustainability of desalination as a remedy to the water crisis in the agriculture sector: an analysis from the climate-water-energy-food nexus perspective *Agric. Water Manag.* **286** 108407

[17] Easley J N, Talozi S and Winter V A G 2023 Feasibility and design of solar-powered electrodialysis reversal desalination systems for agricultural applications in the Middle East and North Africa *Desalination* **561** 116628

[18] Khan Z, Linares P, Rutten M, Parkinson S, Johnson N and García-González J 2018 Spatial and temporal synchronization of water and energy systems: towards a single integrated optimization model for long-term resource planning *Appl. Energy* **210** 499–517

[19] Okampo E J and Nwulu N 2021 Optimisation of renewable energy powered reverse osmosis desalination systems: a state-of-the-art review *Renew. Sustain. Energy Rev.* **140** 110712

[20] Teotónio C, Rodríguez M, Roebeling P and Fortes P 2020 Water competition through the 'water-energy' nexus: assessing the economic impacts of climate change in a Mediterranean context *Energy Econ.* **85** 104539

[21] Chen S and Chen B 2016 Urban energy–water nexus: a network perspective *Appl. Energy* **184** 905–14

[22] Khalkhali M, Westphal K and Mo W 2018 The water-energy nexus at water supply and its implications on the integrated water and energy management *Sci. Total Environ.* **636** 1257–67

[23] Conway D *et al* 2015 Climate and southern Africa's water-energy-food nexus *Nat. Clim. Chang.* **5** 837–46

[24] Ghodsvali M, Krishnamurthy S and de Vries B 2019 Review of transdisciplinary approaches to food-water-energy nexus: a guide towards sustainable development *Environ. Sci. Policy* **101** 266–78

[25] Lee M *et al* 2017 Water-energy nexus for urban water systems: a comparative review on energy intensity and environmental impacts in relation to global water risks *Appl. Energy* **205** 589–601

www.ingramcontent.com/pod-product-compliance
Ingram Content Group UK Ltd.
Pitfield, Milton Keynes, MK11 3LW, UK
UKHW051853100926
8556IPUK00005B/606

9 780750 361